T0140139

Studies in Systems, Decision and Control

Volume 293

The series "Studies in Systems, Decision and Control" (SSDC) covers both new developments and advances, as well as the state of the art, in the various areas of broadly perceived systems, decision making and control–quickly, up to date and with a high quality. The intent is to cover the theory, applications, and perspectives on the state of the art and future developments relevant to systems, decision making, control, complex processes and related areas, as embedded in the fields of engineering, computer science, physics, economics, social and life sciences, as well as the paradigms and methodologies behind them. The series contains monographs, textbooks, lecture notes and edited volumes in systems, decision making and control spanning the areas of Cyber-Physical Systems, Autonomous Systems, Sensor Networks, Control Systems, Energy Systems, Automotive Systems, Biological Systems, Vehicular Networking and Connected Vehicles, Aerospace Systems, Automation, Manufacturing, Smart Grids, Nonlinear Systems, Power Systems, Robotics, Social Systems, Economic Systems and other. Of particular value to both the contributors and the readership are the short publication timeframe and the world-wide distribution and exposure which enable both a wide and rapid dissemination of research output.

** Indexing: The books of this series are submitted to ISI, SCOPUS, DBLP, Ulrichs, MathSciNet, Current Mathematical Publications, Mathematical Reviews, Zentralblatt Math: MetaPress and Springerlink.

More information about this series at http://www.springer.com/series/13304

Sabo Miya Hassan · Rosdiazli Ibrahim ·
Nordin Saad · Kishore Bingi ·
Vijanth Sagayan Asirvadam

Hybrid PID Based Predictive Control Strategies for WirelessHART Networked Control Systems

Springer

Sabo Miya Hassan
Department of Electrical
and Electronics Engineering
Abubakar Tafawa Balewa University
Bauchi, Nigeria

Rosdiazli Ibrahim
Department of Electrical
and Electronic Engineering
Universiti Teknologi PETRONAS
Seri Iskandar, Perak, Malaysia

Nordin Saad
Department of Electrical
and Electronic Engineering
Universiti Teknologi PETRONAS
Seri Iskandar, Perak, Malaysia

Kishore Bingi
Department of Electrical
and Electronic Engineering
Universiti Teknologi PETRONAS
Seri Iskandar, Perak, Malaysia

Vijanth Sagayan Asirvadam
Department of Electrical
and Electronic Engineering
Universiti Teknologi PETRONAS
Seri Iskandar, Perak, Malaysia

ISSN 2198-4182 ISSN 2198-4190 (electronic)
Studies in Systems, Decision and Control
ISBN 978-3-030-47739-4 ISBN 978-3-030-47737-0 (eBook)
https://doi.org/10.1007/978-3-030-47737-0

This Springer imprint is published by the registered company Springer Nature Switzerland AG
The registered company address is: Gewerbestrasse 11, 6330 Cham, Switzerland

This work is dedicated to:

My mother, Fatima, my beloved wife, Amina and daughters Amatullah and Aishatu for their support and love.

—Sabo Miya Hassan

My beloved wife and best friend, Lidia, and my princesses Azra, Auni, Ahna and Ayla, my love will always be with you.

—Rosdiazli Ibrahim

My much-loved family.

—Nordin Saad

My parents Narasimharao and Sudharani, wife Sai Ramya Sri and brother Subbu for their encouragement, care, love and support.

—Kishore Bingi

My parents, my wife Leah and my sons Ahil and Ashanth for their unfailing support.

—Vijanth Sagayan Asirvadam

Preface

This work aims to design and develop an improved PID based control strategies for application in wireless networked environment. The control strategies are expected to permit the peculiarities such as stochastic network delay and uncertainties associated with wireless networked control systems. The strategies will be based on set-point weighting, adaptive set-point weighting, predictive PI and fuzzy PID in order to achieve adequate set-point tracking and disturbance rejection through smoother control action. Likewise, optimisation algorithms based on APSO, BFA and SDA have been proposed to tune the proposed controller approaches. Furthermore, the performance of the proposed strategies has been evaluated on real-time pilot plant for set-point tracking, disturbance rejection and smoother control action.

Therefore, with a total of 6 chapters, the book is structured in such a way to maintain sequential flow for the benefit of readers. Consequently, in Chap. 1, an introduction about what is expected in the book is provided. Chapter 2 proposes an adaptation of the PPI controller to wireless networked control system as well a Filtered PPI controller structure that can be used even in the presence of high frequency noise and stochastic network delays. Chapter 3 presents the development of SW and adaptive SW controller for WHNCS. The SW technique is a powerful and simple method based on the feed-forward strategy. The advantage of this technique is that it can be employed to improve systems performance with respect to set-point tracking ability and disturbance rejection capability. The chapter will also use fuzzy tuner to adapt the developed controller to wider variation of the network delay. Chapter 4 proposes two hybrid algorithms synergizing the social ability of the APSO and the exploitative ability of both spiral dynamic algorithm (SDA) and Adaptive SDA (ASDA). In the later part of the chapter, the proposed algorithms will be used to tune FPPI. Chapter 5 proposes improvement to the adaptation of bacterial foraging algorithm (BFA) and its hybridization with accelerated particle swarm optimization (APSO) in order to accelerate its convergence. In the proposed algorithm, the random walk in the chemotaxis stage of the ABFA is updated through the velocity equation of the APSO. This in turn accelerates the convergence of the algorithm. The algorithm will be validated using selected benchmark functions. Subsequently, an optimal fuzzy PID controller for application in a WirelessHART networked control environment characterized by stochastic network delay will be designed using the proposed algorithm. Finally, Chap. 6 presents comparison of the performance of the proposed control approaches for both simulation and practical applications. Here, three of the proposed controllers namely, SW, FPPI and Fuzzy PID are compared alongside conventional PI and Smith Predictor controllers.

We would like to thank Dr. Thomas Ditzinger, Editorial Director of Interdisciplinary and Applied Sciences Engineering and Prof. Dr. Janusz Kacprzyk, Series Editor of Studies in Systems, Decision and Control for acceptance of the publication of this book with Springer. We also want to thank the production team of Springer for their professional and kind assistance.

The book was supported by Universiti Teknologi PETRONAS through the Award of Yayasan Universiti Teknologi PETRONAS Fundamental Research Grants with Nos 015LC0-053, 0153LC0-045 and 0153AA-H16.

Last but not the least, the authors would like to express our gratitude to Dr. Irraivan Elamvazuthi, Dr. Nurlidia, Dr. Tran Duc and Dr. Hassan Buhari for their encouragement in the research.

Bauchi, Nigeria — Sabo Miya Hassan
Seri Iskandar, Malaysia — Rosdiazli Ibrahim
Seri Iskandar, Malaysia — Nordin Saad
Seri Iskandar, Malaysia — Kishore Bingi
Seri Iskandar, Malaysia — Vijanth Sagayan Asirvadam
January 2020

Acknowledgements

This monograph is based on series of articles and conference proceedings that have been published during the first author's Ph.D. study as well as other researches conducted in the same area. Thus, it is essential for us to reuse some works that have been previously published in various papers and publications. However, in many cases, the work has been modified and rewritten for this monograph. Copyright permission from several publishers is acknowledged as follows.

- S. M. Hassan, R. Ibrahim, N. Saad, V. S. Asirvadam and K. Bingi: Adopting Setpoint Weighting Strategy for WirelessHART Networked Control Systems Characterised by Stochastic Delay, IEEE Access, 5, 25885–25896, 2017.
- S. M. Hassan, R. Ibrahim, N. Saad, V. S. Asirvadam, K. Bingi and T. D. Chung: Fuzzy Adaptive Setpoint Weighting Controller for WirelessHART Networked Control Systems. Wireless Sensor Networks—Insights and Innovations, Edited by Philip Sallis, 10/2017: chapter 9: pages 157–175; INTECH., ISBN: 978-953-51-3562-3, DOI: 10.5772/intechopen.70179.
- S. M. Hassan, R. Ibrahim, N. Saad, V. S. Asirvadam and T. D. Chung: A Filtered PPI Controller for WirelessHART Networked Control System, Journal of Control Engineering and Applied Informatics (CEAI), 19(4), 13–24, 2017.
- S. M. Hassan, R. Ibrahim, N. Saad, V. S. Asirvadam, K. Bingi and T. D. Chung: Hybrid ABF-APSO Algorithm with Application to Tuning of Fuzzy PID Controller for WirelessHART Networked Control System, Intelligent Automation and Soft Computing, Online Article, 1–13, 2019.

- S. M. Hassan, R. Ibrahim, N. Saad, V. S. Asirvadam, K. Bingi: Hybrid Accelerated PSO-Spiral Dynamic Algorithms With Application to Tuning of FPPI Controller in a WirelessHART Environment, Journal of Intelligent & Fuzzy Systems, 37(1), 597–610, 2019.

Bauchi, Nigeria — Sabo Miya Hassan
Seri Iskandar, Malaysia — Rosdiazli Ibrahim
Seri Iskandar, Malaysia — Nordin Saad
Seri Iskandar, Malaysia — Kishore Bingi
Seri Iskandar, Malaysia — Vijanth Sagayan Asirvadam
January 2020

Contents

Acronyms

2DoF	Two-Degree-of-Freedom
ABFA	Adaptive Bacterial Foraging Algorithm
APSO	Accelerated Particle Swarm Optimisation
ASN	Absolute Slot Number
BMI	Bilinear Matrix Inequality
CRC	Cyclic Redundancy Check
CV	Control Valve
DC	Direct Current
DHCP	Dynamic Host Configuration Protocol
DLPDU	Data Link Protocol Data Unit
DO	Disturbance Observer
DSSS	Direct Sequence Spread Spectrum
DTC	Deadtime Compensator
EDDL	Electronic Device Description Language
FHSS	Frequency Hopping Spread Spectrum
FLC	Fuzzy Logic Controller
FODT	First Order Dead Time
FT	Flow Transmitter
GUI	Graphical User Interface
HABF	Hybrid Adaptive Bacterial Foraging
HART	Highway Addressable Remote Transducer Protocol
IEC	International Electrontechnical Commission
IMC	Internal Model Control
IP	Internet Protocol
ISA	International Society of Automation
ISM	Industrial Scientific and Medical
ITAE	Integral Time Absolute Error
IWSN	Industrial Wireless Sensor Networks
LED	Light Emitting Diode
LOS	Line of Sight

LS	Level Sensor
MAC	Media Access Control
MIC	Message Integrity Code
MPC	Model Predictive Controller
MV	Manipulated Variable
NM	Network Manager
OSI	Open Systems Interconnection
PAN	Personal Area Network
PDO	Predictive Disturbance Observer
PID	Proportional Integral and Derivative
PPI	Predictive PI controller
PSO	Particle Swarm Optimisation
PV	Process Variable
RAM	Random Access Memory
RSSI	Received Signal Strength Indicator
SODT	Second Order Dead Time
TDMA	Time Division Multiple Access
WHNCS	WirelessHART Networked Control System
WIA-PA	Wireless networks for Industrial Automation Process Automation
WINA	Wireless Industrial Networking Alliance
WISA	Wireless Interface for Sensor Actuatore
WLAN	Wireless Local Area Network
WSN	Wireless Sensor Network

Chapter 1
Introduction

1.1 Introduction

Wireless sensor networks (WSNs) have gained wide acceptance in the industry especially with the recent approval of three industrial wireless network standards WIA-PA, WirelessHART and ISA100 wireless. This level of acceptance is mainly due to the advantages these standards offer towards industrial process monitoring and control applications over their wired counterpart. Some advantages include: Elimination of cumbersome cabling, the extension of networks capability to areas where it is difficult to run cables, increased data reliability and improve security. Of these three standards, WirelessHART being based on traditional HART protocol has already gained wide patronage in the industry. Thus, in this chapter, the evolution HART protocol as well as the emergence of the WirelessHART technology will first be discussed. Subsequently, the WirelessHART will be compared against its competitors in the industry. Other things to be discussed in the chapter include the architecture of the WirelessHART. Furthermore, the technology will be reviewed for its application in both simulation and practical environments. The later part of the chapter will highlight the purpose and content of the book and lastly a summary will be provided.

1.1.1 Evolution of HART Technology

The evolution of electronics communication between field devices and control systems in the process and automation industry began with the 4–20 mA analogue communication. Then, hybrid systems such as Highway Addressable Remote Transducer (HART) protocol were introduced. These systems combine both the analog and digital signals by modulating an FSK signal over the 4–20 mA current loops, thus permitting the coexistence of the two systems [1, 2]. At a later stage, digital communication technologies such as FOUNDATION™, Fieldbus and PROFIBUS were introduced. Finally, the evolution of industrial communication leads to wireless technologies

S. M. Hassan et al., *Hybrid PID Based Predictive Control Strategies for WirelessHART Networked Control Systems*, Studies in Systems, Decision and Control 293,
https://doi.org/10.1007/978-3-030-47737-0_1

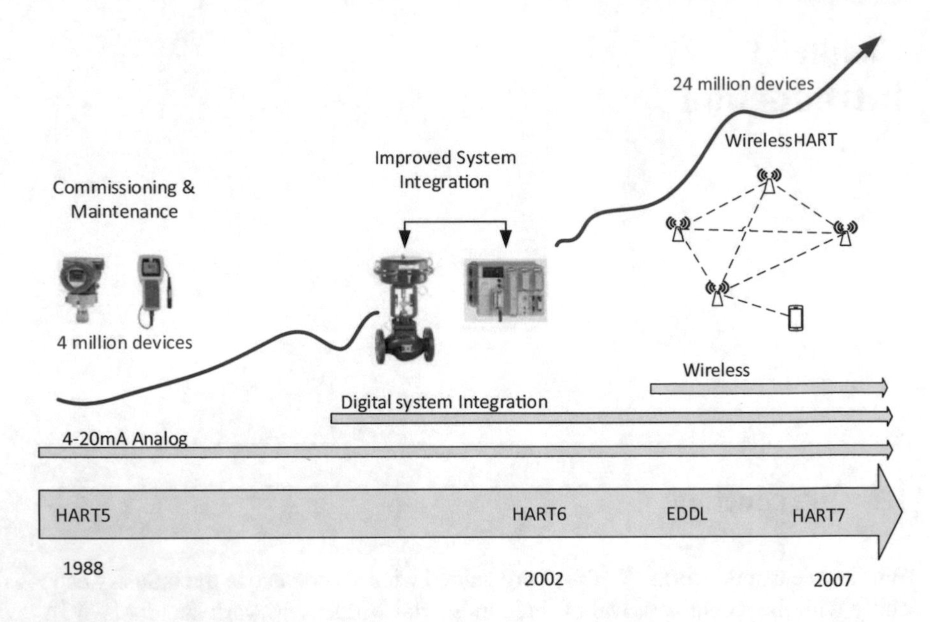

Fig. 1.1 Evolution of HART protocol

such as WirelessHART, WIA-PA and ISA100.11a standards [3–6]. Since the late 80s, the HART protocol has also evolved from being a simple 4–20 mA based hybrid protocol to a sophisticated wired and wireless one. The recent HART protocols have improved on the earlier versions by adding features like enhanced security, event notification, advanced diagnostics, block mode transfer etc.

Evolution of the HART protocol can be described diagrammatically in Fig. 1.1. From the figure, it can be seen that from 1988, with only around 4 million wired devices, the standard has incorporated devices such as digital control valves and controllers with its HART6 in 2002. By 2007, the EDDL and wireless technology have already been integrated into its latest version of the HART protocol (HART7). Presently, there are more than 30 million HART devices installed globally and it is projected that this figure will reach over 46 million by 2021 [7]. The latest version 7 of the HART protocol, introduces the WirelessHART standard by adding the following:

- Mesh wireless networking
- Time synchronization and stamping
- Transport layer
- Network layer
- Security encryption and decryption
- Enhanced burst mode messaging and
- Pipes for high speed file transfer.

1.1.2 *Emergence of WirelessHART Technology*

The importance of wireless technology is becoming more glaring in both the public and industrial sectors [8–10]. Wireless technology employed in the wireless sensor networks (WSNs) is one of the most demanded in the industry and is guaranteed to provide the same or even better control services than its wired counterpart [1, 2, 11–14]. In addition, the WSNs have several advantages over their wired network counterparts [11, 15]. Firstly, the technology eliminates the limitation associated with costly and cumbersome cabling [12, 16]. This will, in turn, eliminate cable maintenance and greatly reduce deployment, redeployment, installation and commissioning times of sensor nodes in the network. Secondly, the capability of wired networks can be extended to such areas where cables are difficult or cannot reach (i.e. environments considered dangerous to run cables) [1]. Another advantage of wireless sensor networks is that, unlike their wired counterparts, they are self-organizing [13] with the ability to support many battery-powered nodes [17–19].

However, concerns about security, reliability, safety, device interoperability and integrity have caused great delay in the acceptance, adoption, and deployment of the WSNs in the industry [14, 20, 21]. The reason is that none of the available wireless technologies is matured enough to provide for real-time performance [22]. Another reason for the slow acceptance and lack of wide adoption of any of the wireless technologies is the absence of an open standard that will ensure customers are not tied to a single supplier and also meets the stringent requirements of the industry [8, 17, 18, 23]. With the coming on board of such open standards, the benefit of wireless technology will dominate the risks posed by uncertainties in deploying the technology in the industry [24, 25].

Several Industrial Organizations such as HART, ISA, WINA and ZigBee have been actively working on improving the application of wireless technologies in industrial automation. As a milestone of such efforts, the HART communication foundation released the Version 7 of the HART protocol and ratified the WirelessHART in 2007 [24, 26, 27]. WirelessHART is the first complete interoperable and open wireless sensor network standard, specifically designed for process monitoring and control applications [8]. WirelessHART assured to maintain the tradition of simplicity and robustness known to users of the earlier versions of the HART protocol. The mesh topology structure of network allows for the possibility of each device in the network to be used as a router to neighbouring devices, thereby creating redundant routes and extending the range of the network. In case of any incidence of obstruction, interference or interruption in a given route, the self-organizing network simply reroutes the communication to another possible route in the mesh network. This feature of the WirelessHART network ensures increased reliability. In addition, the new standard is based on the HART protocol which has about 30 million devices already in operation and it is the most widely used communication protocol in the industry [26, 28]. Hence, there will be very little or no need for training the plant operators to start using the WirelessHART.

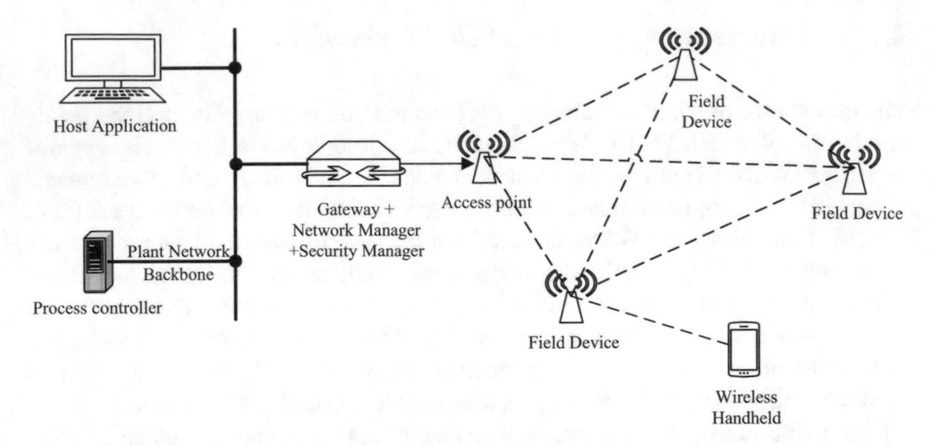

Fig. 1.2 Typical WirelessHART network

The application of modulation methods of frequency-hopping spread-spectrum (FHSS) and direct-sequence spread-spectrum (DSSS) and the use of spatial path diversity and retransmission capability of the mesh network ensure high expectation of robust communication in the system. The standard has also taken care of the issue of data security, ensuring that the users have the choice of selecting the level of security required for their plant. This was made possible through the adoption of a multi-layered technique for data authentication and the use of well-tested encryption algorithms for encryption [26]. The Structure of the WirelessHART network is shown in Fig. 1.2. The WirelessHART network consist of primarily five basic elements which include:

1. Field Devices that are attached to the plant process,
2. Wireless hand-held used for device configuration, diagnostics and calibration,
3. A gateway that connects host applications with field devices,
4. A network manager responsible for network configuration, scheduling and communication management between WirelessHART devices, and
5. A security manager that manages and allocates security encryption keys, and also keep track of devices approved to join the network [11, 23].

1.2 WirelessHART Versus Other Wireless Technologies

WirelessHART is an improvement to the wired HART standard that offers a relatively low speed (e.g. compared to IEEE 802.11g) and cheaper wireless connection. Just like most of the communication standards devised for industrial application, the WirelessHART is based on the Open Systems Interconnection (OSI) model and it adopts the IEEE 802.15.4-2006 for its physical layer. Moreover, it operates in the near globally available unrestricted 2.4 GHz Industrial Scientific and Medical (ISM) radio frequency band using 15 different channels (11–26). A key difference between WirelessHART and other similar standards like ZigBee and WIA-PA is that it specifies its MAC layer which is time-synchronized. The media access control (MAC) header is designed to support co-existence with other networks such as WIA-PA, ISA100 wireless, ZigBee, Wi-Fi, WiMax, etc. Additionally, communication between devices is accomplished using Time Division Multiple Access (TDMA) with strict 10 ms time slots in a super-frame.

Other features of the WirelessHART include channel hopping to evade interference's and minimize multi-path fading effects, channel blacklisting and for the security of the network, it employs the use of industry-standard AES-128 ciphers and keys. Discussions and analysis on the security features of the standard have been conducted [11, 29] and it was revealed that despite some limitations due to its wireless nature, the standard is strong enough security wise to be used in the industrial process control environment [25, 29]. The self-organizing and self-healing mesh networking nature of the WirelessHART is supported by the network layer. Through graph routing and source routing, as monitored and controlled by the network manager, efficient and uninterrupted communication is ensured between devices [30].

Although Wi-Fi standard operates on the same 2.4 GHz unrestricted ISM radio frequency band as the WirelessHART, the two operate on two different standards of IEEE 802.11 and IEEE 802.15.4 for the former and latter respectively. However, Wi-Fi is targeted at WLAN, it is less secured, consumes a lot of power and uses only one channel hence does not support channel hopping like WirelessHART that uses 15 channels [1]. Furthermore, Wi-Fi supports star topology as against WirelessHART mesh (and or star, cluster) [13] topology thus making the Wi-Fi unreliable and therefore unsuitable for the industrial environment as well.

The Bluetooth was ratified as IEEE802.15.1 standard in 2002 as against WirelessHART standard of IEEE 802.15.4 [31]. The technology targets mainly Personal Area Networks (PAN) with a range of up to 60 m. Although both technologies support time slots and channel hopping, Bluetooth only supports star type network topology, and one master support a maximum of 7 slaves only [8]. The restriction imposed by the size of the Bluetooth network makes it impractical to be used in large industrial automated systems. Additionally, the constrained of being a star topology only network makes it not robust enough and therefore highly unreliable. On the contrary, the topology of a WirelessHART network can take the form of a star, a cluster or a mesh, thereby allowing for better scalability and reliability [8].

Released firstly in 2004 and improved in 2006, the ZigBee standard network operates on a single channel and does not allow for hopping between channels throughout its lifespan. This limitation exposes the network to noise and signals interference [32]. As such the standard is still not suitable for industrial application characterized by harsh environments. In 2007, an attempt to make the standard robust enough for industrial application was made with the introduction of ZigBee PRO into the market. The ZigBee PRO has an added feature of improved security and is specially made to allow for the complete network to change operating channel in the event of poor communication caused by either noise or interference or both. This ability to change the channel is referred to as "frequency agility" [11]. Even with the frequency agility features of the ZigBee PRO, it can still not match the frequency hopping ability of the WirelessHART. The ZigBee Alliance prefers to adopt the IEEE 802.15.4 specification in its entirety; hence not ready to engage in any attempt to modify the IEEE 802.15.4 MAC layer. Modification of layer is a requirement to achieve the frequency hopping ability [4, 32–34].

One of the standards proposed for industrial process automation alongside WirelessHART and ISA100.11a (ISA100 Wireless) is the Wireless networks for Industrial Automation-Process Automation (WIA-PA) which was approved by the International Electrotechnical Commission (IEC) as IEC 62601 in 2011 [6, 35, 36]. The standard is also based on the physical and MAC layer of IEEE802.15.4. The standard supports adaptive frequency hopping and is a Mesh+star topology network [37]. Although this standard has a lot of similarities with the WirelessHART, it is however not popular in the market as there are no confirmed suppliers for the device outside its origin country China. Thus WirelessHART being the first and being based on the traditional HART will clearly take the lead.

On the other hand, the ISA100.11a is the closest competitor to the WirelessHART standard for industrial wireless automation applications. While WirelessHART has been approved by the IEC as a first global wireless standard and designated as IEC 62591 since 2010, the ISA100.11a has been approved with the same status by the IEC late 2014 and is designated as IEC 62734. The two standards adopt a simplified version of the OSI model with some adjustment to some of the protocol layers. The physical layer of both the ISA.100.11a and WirelessHART is also based on the IEEE 802.15.4 standard. They both operate on the 2.4 HGz radio frequency band and on 2 MHz bandwidth, 5 MHz equally spaced channels 11–25 and an additional optional channel 26 for ISA.100.11a only. The two standards both specify their MAC layers in such a way that they employ the TDMA technique and use both the DSSS and the FHSS for modulation. While WirelessHART provides for one frequency hopping scheme and specifies fixed 10 ms time slot, the ISA100.11a allows for up to three channel hopping techniques and a time slot range of 10–14 ms [28, 38]. Another difference between the two standards is in the area of device functionality. In the WirelessHART, all field devices (and/or adapters) acts as routers that can forward and receive data to and from other field devices [28, 39]. Furthermore, they can also enable new devices seeking to join the network hence the preference of the mesh topology structure of the network. On the contrary, devices in the ISA100.11a network are defined based on their roles as either input or output devices. It is on this basis that these devices can be configured as end nodes with or without routing capability. By implication, unlike WirelessHART, not all devices can enable other devices to join the ISA100.11a network except those configured to do so. This informs the decision for the standard to be a star, mesh or star-mesh structure [28].

The WirelessHART has an edge over the ISA100.11a standard when interoperability, flexibility, simplicity and acceptability issues are considered [40]. Furthermore, whereas the WirelessHART is the first to be released and to be approved by International Electrotechnical Commission (IEC) as IEC 62591 standard first in 2010 and revised in 2016, the ISA100.11a standard is approved by the IEC as IEC 62734 first in 2012 and with revision in 2014. As mentioned earlier, there are around 30 million HART devices already installed globally that can easily be compatible with the WirelessHART. This gives it a clear lead for the moment in the Industry than both ISA100.11a and WIA-PA. References [6, 28]. To further confirm this point, authors of [6] reported that by 2014, WirelessHART is the most preferred by the adopters among the industrial wireless sensor network (IWSN). The standard is preferred by 25% of the adopters compared to 11% of the ISA100.11a standard. Other IWSN standards such as WIA-PA, ZigBee, wireless interface for sensors and actuators (WISA) put together are preferred by 28% while Hybrid Strategy is preferred by 16%.

Table 1.1 summarizes the comparison between features of some selected wireless standards. The factors considered here is the application of these standards for industrial monitoring and control. In the table, it can be seen that the three industrial standards WIA-PA, ISA100.11a and WirelessHART share several features in common. These features include security, reliability, scalability, topology and low power consumption. Furthermore, classification of some technologies based on standard, the area of application and type of technology is shown in Fig. 1.3. From the figure, it can be seen that all the industrial standards are based on the physical layer IEEE802.14.5 Standard. However, this physical layer is modified by each standard to suit its application. Table 1.2 gives the limitations of the major wireless standards compared to WirelessHART.

1.2.1 WirelessHART Architecture

The WirelessHART protocol stack utilizes five of the seven layers of the OSI communication model. The five layers as shown alongside the wired HART and the original seven OSI layers in Fig. 1.4 include: the physical layer, the data link layer (or sometimes MAC), the network layer, the transport layer and the application layer. An additional feature of the WirelessHART is the central network manager which generates and manages the routing and also decides the communication schedule. These layers are explained briefly as follows

1.2.1.1 Physical Layer

The physical layer of the WirelessHART is based on the IEEE 802.15.4 standard 2006. The standard operates on the 2.4 GHz globally available license-free ISM radio frequency band. The data rate can be up to 250 kb/s and a 2 MHz bandwidth for each of the 5 MHz equally spaced fifteen channels (11–25) ranging from 2405–2475 MHz. It should be noted that channel 26 is not specified for WirelessHART since it is only available for some countries (optional for ISA.100.11a only). The maximum output power is 10 dBm with a maximum range of 75 m non line of sight (LOS) and up to 200 m LOS [8].

Table 1.1 Comparison between features of some selected wireless standards

Feature	Standard					
	Wi-Fi	Bluetooth	ZigBee	WIA-PA	ISA100.11a	WirelessHART
Security	Low	Optional	High	Very high	Very high	Very high
Reliability	Low	Low	Very low	Very high	Very high	Very high
Power consumption	High	High	High	Low	Low	Low
Scalability	Medium	Limited	Medium	High	High	High
Network topology	Star	Star, point-to-point	Star, tree, or mesh	Mesh+Star	Mesh, star	Star, mesh
Data rate	High (11–105 Mbps)	Medium (1 Mbps)	Low (20–250 kbps)	Low (up to 250 kbs)	Low (up to 250 kbs)	Low (up to 250 kbs)
Channel hopping	No	Yes	No	Yes	Yes	Yes
Peer to peer communication	No	Full/Limited	Full/Limited	Limited	Full/Limited	Full
Manager architecture	Centralized	Centralized	Centralized/Distributed	Centralized/Distributed	Centralized	Centralized
Device routing ability	No	No	Limited	Limited	Full/Limited	Full
Frequency channels	5 (2.4 GHz)	79, 40 (2.4 GHz)	27 (All bands)	16 (2.4 GHz)	16 (2.4 GHz)	15 (2.4 GHz)

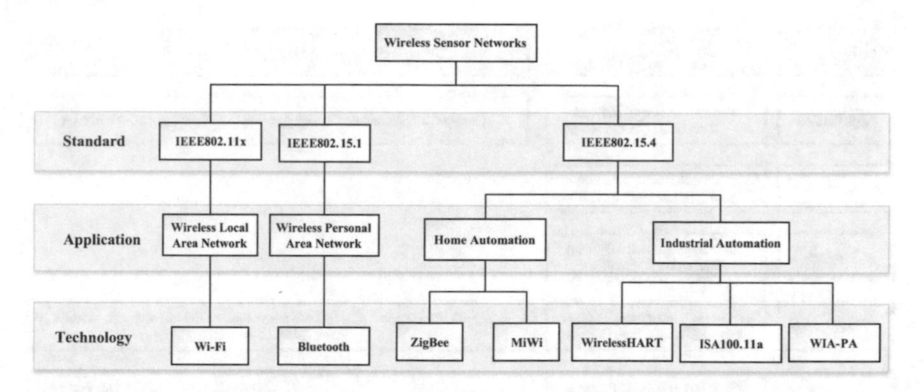

Fig. 1.3 Classification of wireless technology according to standard, application and technology

Table 1.2 Limitation of some wireless standards as compared to WirelessHART

Standard	Target application	Limitation(s)	References
Wi-Fi	WLAN	Less secured, less reliable, not robust, high power consumption, uses only one channel, no frequency hopping, low flexibility, no device routing ability	Muller et al. [1], De Biasi et al. [17, 18]
Bluetooth	PAN	Low reliability, very low scalability (max 8 devices), high power consumption, limited range, very low flexibility	Song et al. [8], [31]
ZigBee	Home automation	Very low reliability and no frequency hopping, low flexibility	Ukarande et al. [34], Lenvall et al. [32]
ISA100.11a	Ind. proc. monitoring and control	Interoperability issues, complex, less accessible than WirelessHART	Petersen et al. [28], Nixon et al. [40]
WIA-PA	Ind. proc. monitoring and control	Limited flexibility, no routing ability, only available in china	Wang and Jiang [6], Liang et al. [37]

1.2.1.2 Data Link (or MAC) Layer

A very distinctive feature of the WirelessHART is its time-synchronized data link layer and the use of TDMA technology to provide for communication that is devoid of interference. This is achieved by grouping a sequence of successive time slots of strictly 10 ms each into a periodic superframe. It is worth noting that the total

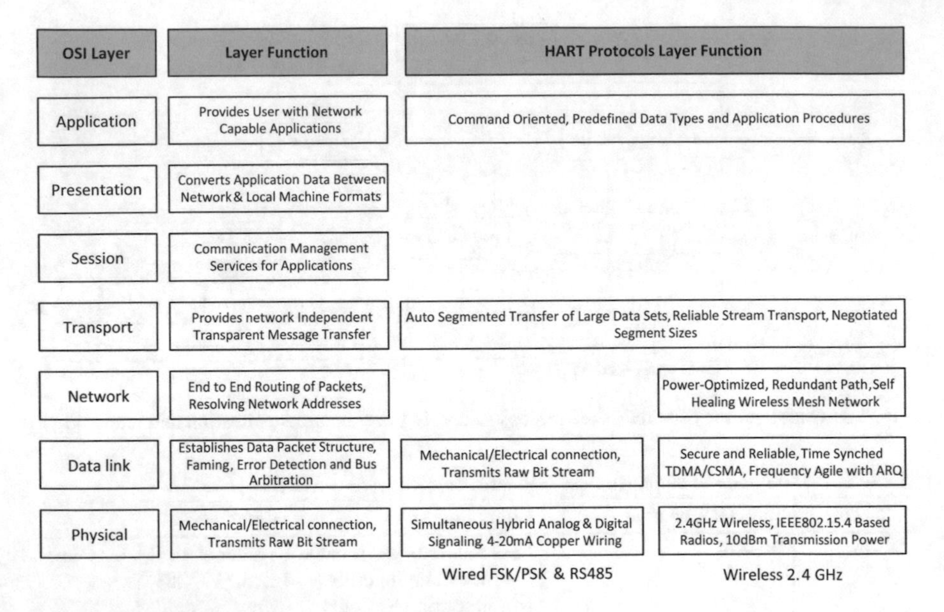

Fig. 1.4 Wired HART and WirelessHART protocol based on OSI layers

lengths of the entire member slots constitutes the period of the superframe itself. The initial point of all the superframes in the network is the time when the network is initially formed and denoted by the actual slot number (ASN) 0, and each superframe is repeated based on its period along the time base. A transaction in time slot in WirelessHART is defined by a vector:

$$\{frame\_id,\ index,\ type,\ src\_addr,\ dst\_addr,\ channel\_offset\}$$

where $frame\_id$ pinpoints a particular superframe; index represent the index of a particular slot in the superframe; type specifies the type of the slot (i.e. transmit, receive or idle slot); src_add and dst_addr are the addresses of the source and destination devices respectively; $channel\_offset$ provides the logical channel to be used in the transaction. Channel blacklisting is employed by the WirelessHART to optimize the channel utilization. Through this channels affected by regular intrusions are blacklisted and the network administrator can disable these channels completely. An active channel table is maintained by each device in order to support channel hopping. The number of entries in the active channel table is affected by blacklisting, thus, a table may have less than 15 entries. The actual channel for a given channel off-set and slot is obtained from the formula given in [8] as follows:

$$ActualChannel = (ChannelOffset + ASN)\%NumChannels \tag{1.1}$$

To obtain the physical channel number, the actual channel number is used as an index into the active channel table. The same channel offset could be mapped to different physical channels in different slots since the ASN is increasing regu-

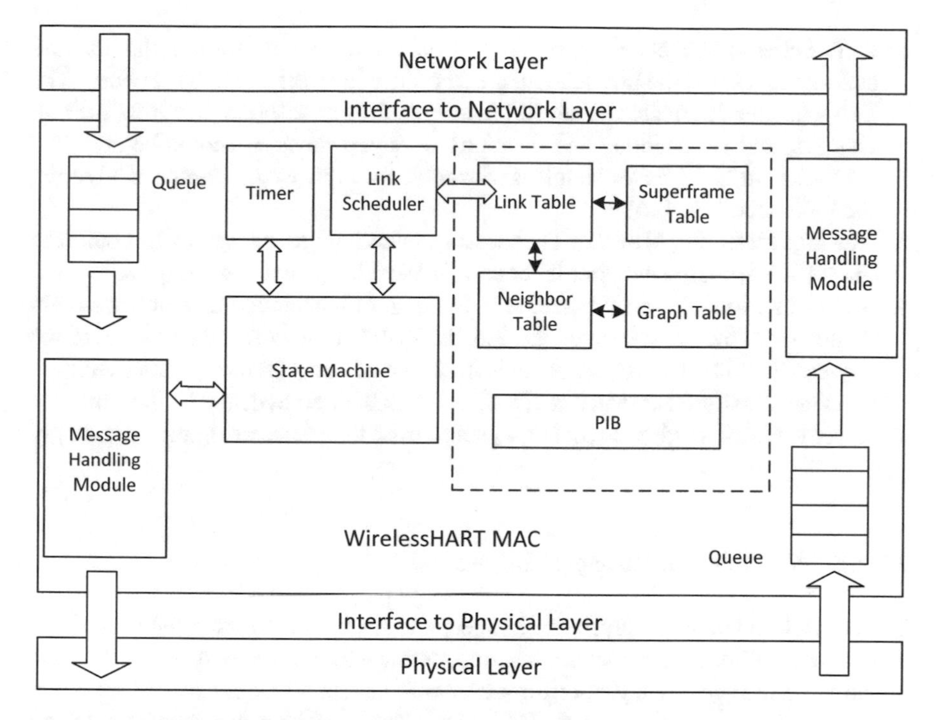

Fig. 1.5 WirelessHART data link architecture

larly. With this, channel diversity is provided and the reliability in communication is improved. A brief description of the overall design of the data dink layer architecture is presented in Fig. 1.5. As seen from the figure, the fundamental sections of the architecture are explained as follows:

1. **Interfaces**: The interface between the MAC and PHY layer defines the service primitives (calling functions) provided to the MAC layer by the physical layer. Likewise the interface between the MAC and NETWORK layer describes the service primitives accorded to the network layer by the MAC layer [26]. Example of such primitives include Connect, Data, Flow Control, and Disconnect and the four basic primitive commands used for communicating data are request, indication, response and confirmation.
2. **Timer**: The timer being an essential unit in the WirelessHART provides precise timing to guarantee proper functioning of the system. The timer ensures the strict 10 ms slot timing requirement and maintains synchronization of the slots [26].
3. **Communication Tables**: A collection of tables is kept by individual network device in the data link layer. Communication configurations generated by the network manager are stored by the superframe and link tables. The list of neighbor nodes that the device can reach directly is kept in the neighbor table while the graph table is used to work together with the network layer and to keep track of routing information [26].

4. **Link Scheduler**: Depending on the schedule of communication in the link and superframe tables, the link scheduler determines the next slot to be serviced. The link scheduler is affected by several factors such as transaction priorities, the link changes, and the enabling and disabling of superframes. It should be note that reassessment of the link schedule is warranted for every event capable of affecting the link scheduling [26].
5. **Message Handling Module**: The message handling module serves as a buffer to the packets from the network layer and physical layer independently [26].
6. **State Machine**: The state machine in the data link layer contains of three main components: the TDMA state machine, the XMIT (transmit) and RECV (receive) engines. The TDMA state machine is in charge of both slot transaction execution and timer clock adjustment. The XMIT and RECV engines deal with the hardware directly, which send and receive a packet over the transceiver correspondingly [26].

1.2.1.3 Network and Transport Layers

The network and transport layers work together to ensure secure and reliable handling of network traffic, communication routing, session formation, and security. In the actual sense, the network layer performs the combined functions of network, transport and session layers, by taking care of the roles of the three layers as demanded by the protocol for those layers in the OSI stack [8, 26]. The overall design of the network and transport layer is shown in Fig. 1.6.

The Network Manager is in charge of forming and maintaining the routing tables of every device to join the network. Three routing techniques are specified in the WirelessHART Network, formation of these routes depends on the communication and performance needs. The three are: (1) Graph routing applied when devices have joined the network and are configured by the network manager. (2) Source Routing, required for diagnostics purpose of the device or part of the network. Here a fixed connection is established between source and destination and (3) Proxy Route, applied when the device has not joined the network yet [8]. The network manager constantly adapts the entire network graph of the network and network schedule to changing network topology and communication demand [8, 26].

1.2.1.4 Application Layer

The application layer is the highest layer of the WirelessHART protocol. Different device commands, responses, data types and status reporting are defined by the application. The layer also handles the message content parsing, extracting command number, and execution of the specified command. The architecture of the application layer is shown in Fig. 1.7.

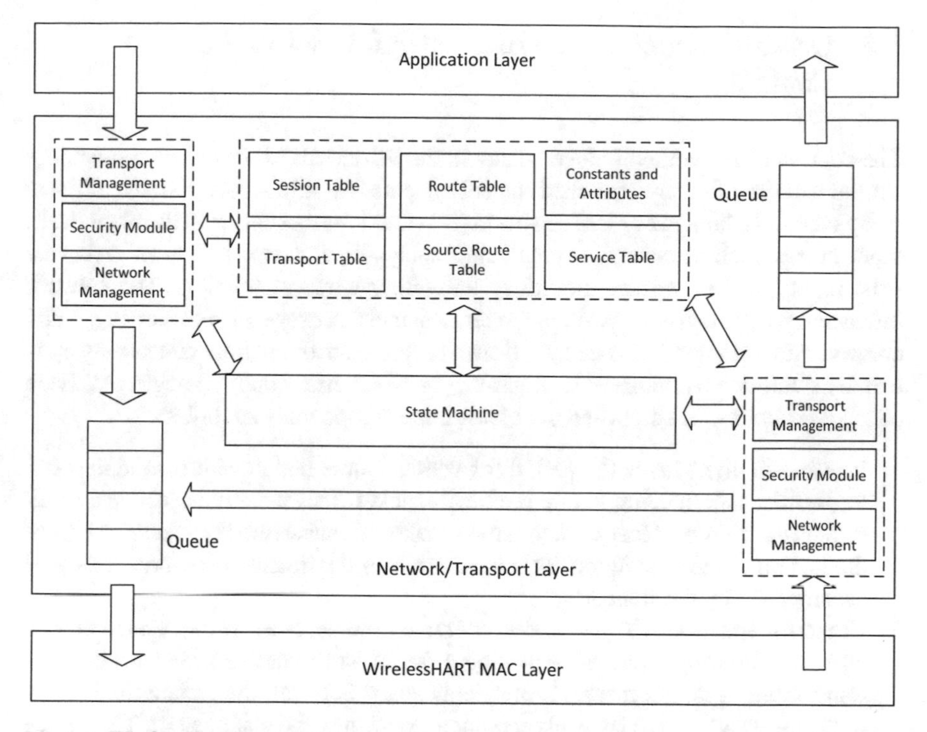

Fig. 1.6 WirelessHART network layer architecture

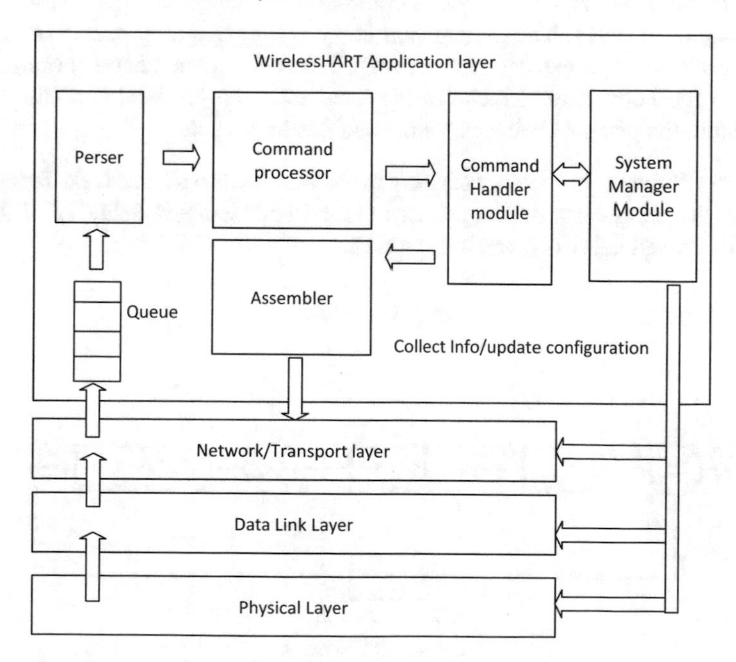

Fig. 1.7 WirelessHART application layer architecture

1.3 General Structure of WirelessHART Networked Control

The sources of networked induced delay in the WirelessHART control network are similar to those of other networked control systems [41–43]. Apart from the intrinsic delay otherwise known as plant deadtime (L_P) that is associated with the plant itself, other delays such as controller processing delays (τ_C), network-induced delay τ_N exist in a typical networked control system either wired or wireless. The network induced delay consists of two components namely sensor-to-controller (τ_{SC}) and controller-to-actuator (τ_{CA}) delays. However, the controller processing delay τ_c is usually small when compared to either τ_{SC} or τ_{CA}, hence could be neglected. Each of the τ_{SC} and τ_{CA} is composed of at least three components as follows:

1. **Waiting delay**: This is the period for which source has to wait for the network availability and queuing before sending a packet. This usually occurs as a result re-transmission of a lost packet, which makes all subsequent packets wait until the lost one is retransmitted. This is inherent in the transmission protocols [44] usually used by the standards.
2. **Transmission delay**: Transmission or frame time delay is the delay encountered when transmitting a packet or placing a frame on the network. For instance, in transmitting a digital message, this delay spans between the instant the first bit of the message left the transmission node to the time the last one left. The source here could be either sensor node or controller.
3. **Propagation delay**: The propagation delay is time taken by packet or frame to propagate through the wireless network. This delay depends on several factors within the network that includes the baud rate, the physical condition of the medium, the distance between source and destination etc.

Figure 1.8 shows the location of delay in the WirelessHART network. To facilitate analysis, the delays can be lumped as total network-induced delay τ_N. Thus, the network-induced delay is given in Eq. (1.2).

$$\tau_N = \tau_{CA} + \tau_{SC}, \tag{1.2}$$

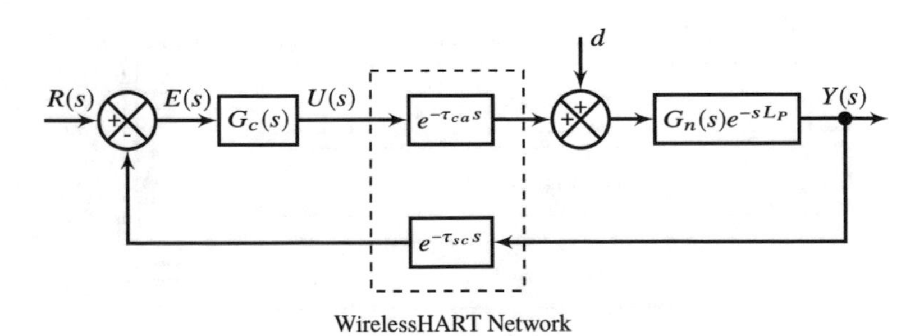

Fig. 1.8 Network delay representation in a single loop WirelessHART networked control system

where τ_{CA} and τ_{SC} are the controller-to-actuator and sensor-to-controller delays respectively, $G(s) = G_n(s)e^{-sL_P}$ is the plant model. It should be noted that in this work, not only the network-induced delay will be considered, but also the process dead-time since it also contributes to the degradation of the control performance of the network. Thus, assuming there is commutativity, the process dead-time (L_P) as given in the figure can be added to the network delay which now gives the total closed loop delay as

$$L = \tau_N + L_P. \tag{1.3}$$

1.3.1 Application of WirelessHART for Control

Despite the wide patronage the WirelessHART has received, its application for control in the industry is faced with challenges. A prominent among these challenges is the network stochastic delay. Thus, good and simple control strategy to handle the challenges of the delay is needed. Although at the lowest control level the PIDs are the most employed, their usage extends to even higher level control [45]. The patronage the PIDs enjoy is mainly due to the key features of simplicity, and ease of tuning manually. The derivative term of the PID is usually turned off due to its sensitivity to noise. Thus, the PI controller is the most commonly used among the PID variants [46]. Even though the derivative action in a PID controller leads to phase advance hence predictive capability (see [47]), it becomes problematic when long dead-time and high-frequency noise are involved. This is due to noise sensitivity introduced by the derivative term.

On the other hand, the predictive mechanism of the PID is also not suitable when considering non-minimum phase. Thus, it is difficult to control processes characterized by long dead-time and variable network delay with standard feedback controllers such as PID. Consequently, PID controllers are inadequate to be employed in a delayed environment or where there are high-frequency noise and other uncertainties. When PID controllers are used in environments such as WirelessHART, the result is oscillation and instability because the controllers are limited in gain [48]. For example, if a too small gain is used, the response of the system becomes sluggish. On the other hand, if a too large gain is used the response becomes oscillatory and could even be unstable. Furthermore, the conventional PID algorithms used in wired control presume that the input and output paths to be reliable and measurements are received periodically. If used in wireless control and the input is temporarily lost, the PID controller will accumulate error based on the last received value causing a spike at the output and a possibility of large process oscillations once the communication is restored. Problems are also inevitable if the output communication is interrupted.

Several researchers have proposed the improvement of PID controller to suit its application in time-delay systems [49–51]. At the advanced level, techniques such as deadtime compensator (DTC), model-based predictive controller (MPC) and generalized predictive controller (GPC) and have been proposed for wired and wireless applications [52–54]. A key and common drawback of these controllers is that of complexity, which makes them difficult to be practically implemented [48]. Another reason for the implementation issue is that some of these controllers require the exact model of the process to be controlled. This is impossible in reality [48].

1.3.2 Simulation Environment

The three categories of applications running in any process plant as defined by ISA [26] (and considering the increasing order of criticality) are for monitoring, control and safety. Wireless technology has been applied for monitoring purpose and attempts are being made to apply it to the control aspect [27, 28, 55]. The WirelessHART is the first open standard that has been proposed for monitoring and control applications in the industry. Most of the work done on WirelessHART so far is based mostly on the development, analysis, performance evaluation of the standard, interoperability test and the implementation in the simulation environment [8, 15, 16, 24–26, 38, 56–72].

The first attempted application of the standard for control purpose started with the True-Time that is, a wireless simulation software specifically designed for simulation of networks supporting IEEE802.11 and ZigBee networks [73]. In the True Time software, the execution of MAC protocol for all devices is done in the wireless network block, since it is based on ZigBee and WLAN. This situation is a deviation from reality. In reality, every device in the network has own sublayer of MAC. To reflect this scenario, of which has been already supported by WirelessHART, the capability of the software was extended. The standard was then simulated for process control subject to two conditions of clock drift and packet losses [17, 18].

A methodology for co-design of communication scheduling and controller for control systems operating over WirelessHART system was proposed aimed to address issues relating to time-optimal convergecast scheduling of data and dissemination, end-to-end reliability subject to packet loss and controller performance [55]. A simulator and a design process support tool targeting WirelessHART for process control were also presented in [74]. The simulator specifically targets efficient communication and processing scheduling, but it must be extended to take care of multi-hop communication and also to take care of all the 15 channels. In [38, 75], A hybrid simulation approach based on COOJA and Contiki operating system to evaluate the performance of WirelessHART network and WirelessHART enabled devices was presented. Another co-simulation technique based on the interaction of True-Time and OMNET++ was proposed. In this work, a simple control loop involving a DC servomotor was used. Results obtained from that simulation framework justifies the suitability of WirelessHART for closed-loop control [76]. A hybrid control oriented model for WirelessHART networked control systems under source routing was proposed in [77]. However, this proposal like the ones mentioned above is far from being practically implemented. The summary of the applications are given in Table 1.3.

Table 1.3 Summary of application of WirelessHART for control: simulation

Publication	Year	Features	Limitations
De Biasi et al. [17, 18], Snicars [73]	2008	True-time extended for WirelessHART, control, subject to clock drift and packet dropout	Does not consider practical network delay and are based on basic PID, DTC
Pesonen et al. [55]	2009	Methodology co-design of comm. scheduling and control with WH	Does not address the issue of stochastic network delay
Shah et al. [74]	2010	Simulator and design process support tool for WirelessHART control	Only a simulator, does not consider real WirelessHART devices
Ferrari et al. [76]	2013	Interaction of true time and, OMNET++ for WH control	Limited to simulation environment and considers only PID

1.3.3 Practical Application

For practical purpose, few applications of the WirelessHART for real-time control were reported. It is reported in [78] that measurements collected over wireless using WirelessHART network used for control purposes can rival those collected using the wired network. Moreover, WirelessHART transmitters were placed alongside wired transmitters and measurements collected on both sides were used for both Column-pressure and heater steam-flow control using a modified PID algorithm for the wireless communication. In [79] it was concluded that control over WirelessHART is achievable after comparing the performance of a simple control loop with that of wired Foundation Fieldbus. Here a simple control loop to control an LED was set up using WirelessHART devices the same way as with Foundation Fieldbus. However, this method only considers star topology thereby not reflecting the mesh nature of the WirelessHART network.

An attempt was also made in [80] to implement WirelessHART for liquid level control for an industrial tank, however, the technology was only used for detecting liquid level not actually control. In [81], three control strategies for WirelessHART networks were proposed, namely the Control in the Host, Control in the Gateway and Control in the Field. However, the WirelessHART only partially supports control in the field, thus, the gateway control was recommended since it is fully supported by the WirelessHART and will have less delay compared to the control implemented in the Host. Here the gateway is modified by adding a function block application layers similar to those in Foundation Fieldbus to facilitate configuration and execution of control modules as shown in Fig. 1.9. Even with this attempt, the control scenario has not yet been applied to a real system.

An improved PID algorithm to take care of slower measurement updates, non-periodic measurements and loss of communication imposed by wireless transmitters

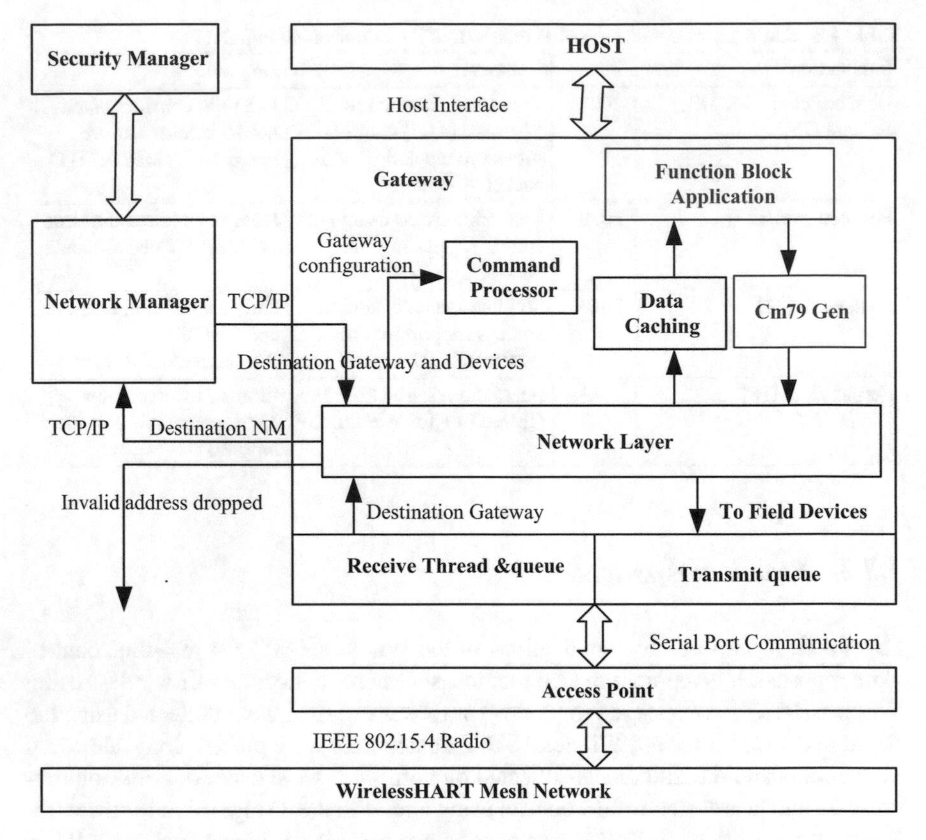

Fig. 1.9 WirelessHART gateway design for control in the gateway

(e.g. WirelessHART transmitters) called PIDPlus algorithm was presented in [52]. Here, two other PID control approaches namely with Kalman filter observer and both modified for wireless measurements were presented and the performances of these two approaches were compared to the performance of the PIDPlus algorithm. It was found that while the Kalman filter observer's performance was better than that of the PIDPlus algorithm, the latter is much better than the in terms of the error. More so, the WirelessHART transmitters were only used for measurement in the feedback loop, the feed-forward loop remained wired. In a related development, measurements collected through WirelessHART transmitters are used for control of dividing wall column in [82] using the method the authors called "Wireless model predictive control". This is similar to the previous case except for the control method used. It is however, worth noting that model predictive controller is difficult and complex to implement especially where delay and mismatches exist.

In [16, 83], the behavior of WirelessHART network is assessed for water level control process. The network delay is measured through the use of time-stamp. The PID control algorithm used in the work is implemented in Java. The results showed

Table 1.4 Summary of application of WirelessHART for control: practical

Publication	Year	Features	Limitations
Han et al. [81]	2010	Strategies, for WH control	Only demonstrates possibility of control with WirelessHART
Morakinyo [80]	2013	Tank liquid level control with WH	Only measurement of liquid level was achieved
Seibert et al. [78]	2011	Control with WirelessHART measurement	Wireless measurement only, PID
Zhu et al. [79]	2012	WirelessHART versus foundation FieldBus	Demonstration of possibility, of wireless control, no real device considered only LED
Blevins et al. [52, 54, 82, 84]	2014, 2015, 2016	Process control with real time WH	Wireless measurement only, PID, MPC
Santos et al. [16], Bertelli et al. [83]	2015, 2017	Assessment, water level control	PID, actuation not fully wireless

that due to the stochastic nature of the network delay, the PID performed poorly with high overshoots and oscillation for all the update rates of 8, 4 and 1 s considered. A relatively better result was obtained for the 1 s period with piping. This however still has some overshoots at step change. A summary of reported works on practical application of WirelessHART for control and their limitations is given in Table 1.4.

In view of the above findings, it can be seen that research into application of WirelessHART is still not very advanced as far as its application for control is concerned. Therefore, research into application of the technology for control especially using the PID, with the view to mitigating the effect of stochastic network delay while overcoming the limitations of the PID with respect to gain limitation is inevitable.

Thus, this book will focus on the development of WirelessHART Networked PID based control strategies for real-time process control to achieve adequate stochastic delay compensation, set-point tracking and disturbance rejection performance through smoother control action. The control methodologies to be developed will be based on set-point weighting, structural modification that includes adaptation and filtering as well as tuning using meta-heuristic algorithms. In this book, an attempt has been made for the real-time implementation of some of the proposed approaches while an extensive simulation and hardware in the loop simulation has been conducted. Another interesting aspect of the book is that hybrid meta-heuristic algorithms have been developed. The algorithms leverage on the speed of convergence of APSO algorithm on one hand and the dynamism of both Bacterial foraging algorithm and spiral synamic algorithm on the other hand.

1.4 Summary

In this chapter, in order to cast mind of the readers on what the WirelessHART and related technologies are all about, a detailed background as well comprehensive comparison between these technologies is provided. Furthermore, the WirelessHART has been reviewed for its application in both simulation and practical environments. Finally, the scope of this book has been provided.

References

1. Muller, I., Netto, J.C., Pereira, C.E.: WirelessHART field devices. IEEE Instrum. Meas. Mag. **14**(6), 20–25 (2011)
2. Muller, I., Pereira, C.E., Netto, J.C., Fabris, E.E., Allgayer, R.: Development of a WirelessHART compatible field device. In: 2010 IEEE Instrumentation and Measurement Technology Conference Proceedings, Austin, Texas, 3–6 May 2010
3. Henk, B.V.D.: Full redundant field wireless automation solutions based on the ISA100.11a standard. Flow Level Press. 14–15 (2014)
4. Habib, G., Haddad, N., El Khoury, R.: Case study: WirelessHART vs ZigBee network. In: 2015 3rd International Conference on Technological Advances in Electrical, Electronics and Computer Engineering, Beirut, Lebanon, 29 April–1 May 2015
5. Flammini, A., Ferrari, P., Marioli, D., Sisinni, E., Taroni, A.: Wired and wireless sensor networks for industrial applications. Microelectron. J. **40**(9), 1322–1336 (2009)
6. Wang, Q., Jiang, J.: Comparative examination on architecture and protocol of industrial wireless sensor network standards. IEEE Commun. Surv. Tutor. **18**(3), 2197–2219 (2016)
7. Olenewa, J.: Guide to Wireless Communications. Cengage Learning, Boston (2013)
8. Song, J., Han, S., Mok, A., Chen, D., Lucas, M., Nixon, M., Pratt, W.: WirelessHART: applying wireless technology in real-time industrial process control. In: 2008 IEEE Real-Time and Embedded Technology and Applications Symposium, St. Louis, Missouri, USA, 22–24 April 2008
9. Willig, A., Matheus, K., Wolisz, A.: Wireless technology in industrial networks. Proc. IEEE **93**(6), 1130–1151 (2005)
10. Chen, D., Nixon, M., Aneweer, T., Shepard, R., Burr, K., Mok, A.K.: Wireless process control products from ISA2004. In: International Workshop on Wireless and Industrial Automation
11. Alcaraz, C., Lopez, J.: A security analysis for wireless sensor mesh networks in highly critical systems. IEEE Trans. Syst. Man Cybern. Part C (Appl. Rev.) **40**(4), 419–428 (2010)
12. Baillieul, J., Antsaklis, P.J.: Control and communication challenges in networked real-time systems. Proc. IEEE **95**(1), 9–28 (2007)
13. Al Agha, K., Bertin, M.H., Dang, T., Guitton, A., Minet, P., Val, T., Viollet, J.B.: Which wireless technology for industrial wireless sensor networks? The development of OCARI technology. IEEE Trans. Ind. Electron. **56**(10), 4266–4278 (2009)
14. Åkerberg, J., Gidlund, M., Björkman, M.: Future research challenges in wireless sensor and actuator networks targeting industrial automation. In: 2011 9th IEEE International Conference on Industrial Informatics, Caparica, Lisbon, Portugal, 26–29 July 2011
15. Nobre, M., Silva, I., Guedes, L.A.: Performance evaluation of WirelessHART networks using a new network simulator 3 module. Comput. Electr. Eng. **41**, 325–341 (2015)
16. Santos, A., Lopes, D., César, J., Luciano, L., Neto, A., Guedes, L.A., Silva, I.: Assessment of WirelessHART networks in closed-loop control system. In: 2015 IEEE International Conference on Industrial Technology (ICIT), pp. 2172–2177 (2015)

17. De Biasi, M., Snickars, C., Landernäs, K., Isaksson, A.J.: Simulation of process control with WirelessHART networks subject to clock drift. In: IEEE 2008 COMPSAC, pp. 1355–1360 (2008)
18. De Biasi, M., Snickars, C., Landernäs, K., Isaksson, A.J.: Simulation of process control with WirelessHART networks subject to packet losses. In: IEEE 2008 CASE, pp. 548–553 (2008)
19. Menezes, M.: WirelessHART® minimizes cost of new energy measurements. Energy Eng. **112**(1), 69–77. Taylor & Francis (2015)
20. Eriksson, J., Österlind, F., Finne, N., Tsiftes, N., Dunkels, A., Voigt, T., Sauter, R., Marrón, P.J.: COOJA/MSPSim: interoperability testing for wireless sensor networks. In: Proceedings of the 2nd International Conference on Simulation Tools and Techniques, 1–27 ICST (Institute for Computer Sciences, Social Informatics and Telecommunications Engineering) (2009)
21. Öztürk, C., Karaboğa, D., Görkemli, B.: Artificial bee colony algorithm for dynamic deployment of wireless sensor networks. Turk. J. Electr. Eng. Comput. Sci. **20**(2), 255–262 (2012). The Scientific and Technological Research Council of Turkey (2012)
22. Zhao, G.: Wireless sensor networks for industrial process monitoring and control: a survey. Netw. Protoc. Algorithms **3**(1), 46–63 (2011)
23. Åkerberg, J., Gidlund, M., Neander, J., Lennvall, T., Björkman, M.: Deterministic downlink transmission in WirelessHART networks enabling wireless control applications. In: IECON 2010-36th Annual Conference on IEEE Industrial Electronics Society, pp. 2120–2125. IEEE (2010)
24. Kim, A.N., Hekland, F., Petersen, S., Doyle, P.: When HART goes wireless: understanding and implementing the WirelessHART standard. In: 2008 IEEE International Conference on Emerging Technologies and Factory Automation, pp. 899–907. IEEE (2008)
25. Nawaz, F., Jeoti, V.: Performance assessment of WirelessHART technology for its implementation in dense reader environment. Computing **98**(3), 257–277. Springer (2016)
26. Chen, D., Nixon, M., Mok, A.: Why WirelessHART, pp. 195–199. Springer, Berlin (2010)
27. Petersen, S., Carlsen, S.: Performance evaluation of WirelessHART for factory automation. In: IEEE Conference on Emerging Technologies and Factory Automation, 2009. ETFA 2009, pp. 1–9. IEEE (2009)
28. Petersen, S., Carlsen, S.: WirelessHART versus ISA100.11a: the format war hits the factory floor. IEEE Ind. Electron. Mag. **5**(4), 23–34. IEEE (2011)
29. Raza, S., Slabbert, A., Voigt, T., Landernäs, K.: Security considerations for the WirelessHART protocol. In: 2009 IEEE Conference on Emerging Technologies and Factory Automation, Palma de Mallorca, Spain, 22–25 September 2009
30. Chen, D., Nixon, M., Mok, A.: Future of wireless and the WirelessHART standard. WirelessHART™. Springer, Boston (2010)
31. Bluetooth, S.I.G.: Bluetooth core specification version 4.0. Specif. Bluetooth Syst. **1**, 7 (2010)
32. Lennvall, T., Svensson, S., Hekland, F.: A comparison of WirelessHART and ZigBee for industrial applications. In: 2008 IEEE International Workshop on Factory Communication Systems, Dresden, Germany, 20–23 May 2008
33. Egan, D.: The emergence of ZigBee in building automation and industrial controls. Comput. Control Eng. **16**(2), 14–19 (2005)
34. Ukarande, V.V., Shaikh, F.I.: WirelessHART a right choice over ZigBee. In: National Conference on Advances in Computing, Networking and Security, Nanded, India, 23–24 December 2013
35. Zheng, M., Liang, W., Yu, H., Xiao, Y.: Performance analysis of the industrial wireless networks standard: WIA-PA. Mob. Netw. Appl. **22**(1), 139–150 (2017)
36. Nobre, M., Silva, I., Guedes, L.: Routing and scheduling algorithms for WirelessHART networks: a survey. Sensors **15**(5), 9703–9740 (2015)
37. Liang, W., Zhang, X., Xiao, Y., Wang, F., Zeng, P., Yu, H.: Survey and experiments of WIA-PA specification of industrial wireless network. Wirel. Commun. Mob. Comput. **11**(8), 1197–1212 (2011)
38. Gustafsson, D.: Wirelesshart-implementation and evaluation on wireless sensors. Masters's Degree Project, KTH University, Electrical Engineering, pp. 1–39 (2009)

39. Akyildiz, I.F., Wang, X., Wang, W.: Wireless mesh networks: a survey. Comput. Netw. **47**(4), 445–487 (2005)
40. Nixon, M., Rock, T.R.: A comparison of WirelessHART and ISA100.11a. Whitepaper, Emerson Process Manag. 1–36 (2012)
41. Tipsuwan, Y., Chow, M.Y.: Control methodologies in networked control systems. Control Eng. Pract. **11**(10), 1099–1111 (2003)
42. Nilsson, J.: Real-time control systems with delays (1998)
43. Hespanha, J.P., Naghshtabrizi, P., Xu, Y.: A survey of recent results in networked control systems. Proc. IEEE **95**(1), 138–162 (2007)
44. Lee, Y.J., Atiquzzaman, M.: Mean waiting delay for web object transfer in wireless SCTP environment. In: 2009 IEEE International Conference on Communications, Dresden, Germany, 14–18 June 2009
45. Ingimundarson, A., Hägglund, T.: Performance comparison between PID and dead-time compensating controllers. J. Process Control **12**(8), 887–895 (2002)
46. Huba, M.: Comparing 2DOF PI and predictive disturbance observer based filtered PI control. J. Process Control **23**(10), 1379–1400 (2013)
47. Larsson, P., Hägglund, T.: Comparison between robust PID and predictive PI controllers with constrained control signal noise sensitivity. IFAC Proc. Vol. **45**(3), 175–180 (2012)
48. Tan, K.K., Tang, K.Z., Su, Y., Lee, T.H., Hang, C.C.: Deadtime compensation via setpoint variation. J. Process Control **20**(7), 848–859 (2010)
49. Åström, K.J., Hägglund, T.: Advanced PID Control. ISA-The Instrumentation, Systems, and Automation Society, Research Triangle Park (2006)
50. Åström, K.J., Hägglund, T.: PID Controllers: Theory, Design, and Tuning. Instrument Society of America, Research Triangle Park (1995)
51. O'Dwyer, A.: Handbook of PI and PID Controller Tuning Rules. World Scientific, Singapore (2006)
52. Blevins, T., Nixon, M., Wojsznis, W.: PID control using wireless measurements. In: 2014 American Control Conference, Portland, OR, 4–6 June 2014
53. Blevins, T., Chen, D., Nixon, M., Wojsznis, W.: Wireless Control Foundation: Continuous and Discrete Control for the Process Industry. International Society of Automation, Research Triangle Park (2015)
54. Blevins, T., Nixon, M., Wojsznis, W.: Event based control applied to wireless throttling valves. In: 2015 International Conference on Event-Based Control, Communication, and Signal Processing, Krakow, Poland, 17–19 June 2015
55. Pesonen, J., Zhang, H., Soldati, P., Johansson, M.: Methodology and tools for controller-networking codesign in WirelessHART. In: 2009 IEEE Conference on Emerging Technologies and Factory Automation, Palma de Mallorca, Spain, 22–25 September 2009
56. Han, S., Song, J., Zhu, X., Mok, A.K., Chen, D., Nixon, M., Gondhalekar, V.: Wi-HTest: compliance test suite for diagnosing devices in real-time WirelessHART network. In: 2009 15th IEEE Real-Time and Embedded Technology and Applications Symposium, Beijing, China, 24–26 August 2009
57. Nobre, M., Silva, I., Guedes, L.A., Portugal, P.: Towards a WirelessHART module for the ns-3 simulator. In: 2010 IEEE 15th Conference on Emerging Technologies and Factory Automation, Bilbao, Spain, 14–16 September 2010
58. Ferrari, P., Flammini, A., Rinaldi, S., Sisinni, E.: Performance assessment of a WirelessHART network in a real-world testbed. In: 2012 IEEE International Instrumentation and Measurement Technology Conference Proceedings, Graz, Austria, 13–16 May 2012
59. Amidi, S., Gandhi, A: An open, standard-based wireless network: connecting WirelessHART® sensor networks to Experion™ PKS using Honeywell's OneWireless™ network. Honeywell (2012)
60. Fiore, G., Ercoli, V., Isaksson, A.J., Landernäs, K., Di Benedetto, M.D.: Multihop multi-channel scheduling for wireless control in WirelessHART networks. In: 2009 IEEE Conference on Emerging Technologies and Factory Automation, Palma de Mallorca, Spain, 22–25 September 2009

61. Trikaliotis, S., Gnad, A.: Mapping WirelessHART into PROFINET and PROFIBUS fieldbusses. In: 2009 IEEE Conference on Emerging Technologies and Factory Automation, Palma de Mallorca, Spain, 22–25 September 2009
62. Zhang, H., Soldati, P., Johansson, M.: Optimal link scheduling and channel assignment for convergecast in linear WirelessHART networks. In: 2009 7th International Symposium on Modeling and Optimization in Mobile, Ad Hoc, and Wireless Networks, Seoul, South Korea, 23–27 June 2009
63. Åkerberg, J., Gidlund, M., Lennvall, T., Neander, J., Björkman, M.: Integration of WirelessHART networks in distributed control systems using PROFINET IO. In: 2010 8th IEEE International Conference on Industrial Informatics, Osaka, Japan, 13–16 July 2010
64. Depari, A., Ferrari, P., Flammini, A., Lancellotti, M., Marioli, D., Rinaldi, S., Sisinni, E.: Design and performance evaluation of a distributed WirelessHART sniffer based on IEEE1588. In: 2009 International Symposium on Precision Clock Synchronization for Measurement, Control and Communication, Brescia, Italy, 12–16 October 2009
65. Raza, S., Voigt, T., Slabbert, A., Landernas, K.: Design and implementation of a security manager for WirelessHART networks. In: 2009 IEEE 6th International Conference on Mobile Adhoc and Sensor Systems, Macau, China, 12–15 October 2009
66. Silva, I., Guedes, L.A., Portugal, P., Vasques, F.: Dependability evaluation of WirelessHART best practices. In: Proceedings of 2012 IEEE 17th International Conference on Emerging Technologies and Factory Automation, Krakow, Poland, 17–21 September 2012
67. Song, J., Han, S., Zhu, X., Mok, A.K., Chen, D., Nixon, M.: A complete WirelessHART network. In: Proceedings of the 6th ACM Conference on Embedded Network Sensor Systems, Raleigh, NC, USA, 5–7 November 2008
68. Ferrari, P., Flammini, A., Marioli, D., Rinaldi, S., Sisinni, E.: On the implementation and performance assessment of a WirelessHART distributed packet analyzer. IEEE Trans. Instrum. Meas. **59**(5), 1342–1352 (2010)
69. Saifullah, A., Xu, Y., Lu, C., Chen, Y.: Real-time scheduling for WirelessHART networks. In: 2010 31st IEEE Real-Time Systems Symposium, San Diego, California, USA, 30 November–3 December 2010
70. Huang, Q., Sikora, A., Groza, V.F., Zand, P.: Simulation and analysis of WirelessHART nodes for real-time actuator application. In: 2014 IEEE International Instrumentation and Measurement Technology Conference (I2MTC) Proceedings, Montevideo, Uruguay, 12–15 May 2014
71. Winter, J.M., Kunzel, G., Muller, I., Pereira, C.E., Netto, J.C.: Study of routing mechanisms in a WirelessHART network. In: 2013 IEEE International Conference on Industrial Technology, Cape Town, South Africa, 25–28 February 2013
72. Müller, I., Winter, J.M., Pereira, C.E., Netto, J.C., Eckard, D.: Automatic RF power adjustment for WirelessHART field devices. In: 2014 IEEE International Conference on Industrial Technology, Busan, South Korea, 26 February–1 March 2014
73. Snickars, C.: Design of a WirelessHART simulator for studying delay compensation in networked control systems, Masters Degree Project, KTH Electrical Engineering, Stockholm, Sweden, Citeseer (2008)
74. Shah, K., Seceleanu, T., Gidlund, M.: Design and implementation of a WirelessHART simulator for process control. In: International Symposium on Industrial Embedded System, Italy, 7–9 July 2010
75. Konovalov, I.: A framework for WirelessHART simulations. Swedish Institute of Computer Science (2010)
76. Ferrari, P., Flammini, A., Rizzi, M., Sisinni, E.: Improving simulation of wireless networked control systems based on WirelessHART. Comput. Stand. Interfaces **35**(6), 605–615. Elsevier (2013)
77. Maass, A.I., Nešić, D., Dower, P.M.: A hybrid model of networked control systems implemented on WirelessHART networks under source routing configuration. In: 2016 Australian Control Conference, Newcastle, Australia, 60–65 November 2016
78. Seibert, F.: Wireless HART successfully handles control. Chem. Process. (2011)

79. Zhu, X., Lin, T., Han, S., Mok, A., Chen, D., Nixon, M., Rotvold, E.: Measuring WirelessHART against wired fieldbus for control. In: IEEE 10th International Conference on Industrial Informatics, Beijing, China, 25–27 July 2012
80. Morakinyo, S.: Implementation of WirelessHART technology for liquid level control and monitoring (2013)
81. Han, S., Zhu, X., Aloysius, K.M., Nixon, M., Blevins, T., Chen, D.: Control over WirelessHART network. In: IECON 2010-36th Annual Conference on IEEE Industrial Electronics Society, Glendale-USA, 7–10 November 2010
82. Blevins, T., Wojsznis, W.K., Nixon, M.J., Roach, B.: Wireless model predictive control applied for dividing wall column control. In: 2016 Second International Conference on Event-Based Control, Communication, and Signal Processing, Krakow, Poland, 13–15 June 2016
83. Bertelli, G., Santos, A., Silva, I., Fernandes, R., Brandao, D., Muller, I., Pereira, C.E.: Research activities on industrial wireless instrumentation: Brazilian perspective. IEEE Instrum. Meas. Mag. **20**(2), 21–30 (2017)
84. Blevins, T., Chen, D., Han, S., Nixon, M., Wojsznis, W.: Process control over real-time wireless sensor and actuator networks. In: 2015 IEEE 17th International Conference on High Performance Computing and Communications, 2015 IEEE 7th International Symposium on Cyberspace Safety and Security, and 2015 IEEE 12th International Conference on Embedded: Software and Systems, New York, USA, 24–26 August 2015

Part I
Hybrid PID Based WirelessHART Networked Control Strategies

Chapter 2
Filtered Predictive PI Controller for WirelessHART Networked Systems

2.1 Introduction

Recently, increasing attention has been paid towards applying wireless technology for control. This is due to its advantages of flexibility, scalability, use of fewer cables and overall reduced operational cost compared to its wired counterpart. However, the technology is often affected by stochastic delay and high frequency noise. PIDs are ill-equipped to deal with these problems while model based controllers such as dead-time compensators (DTCs) like Smith predictor and internal model controllers (IMCs) are complex and require exact plant model for implementation. Thus, predictive PI (PPI) controller being a settlement between the PIDs and the model based controllers is a good candidate. The PPI retains simplicity of the PID, it has the ability to predict long time delay and can be used even with model mismatch. However, the PPI is severely affected by high frequency noise. Therefore, this chapter proposes an adaptation of the PPI controller to wireless networked control system as well a Filtered PPI controller structure that can be used even in the presence of high frequency noise and stochastic network delays. Simulation and experimental results proved the viability of the proposed method.

2.2 Review of Some DTC Techniques

To improve the performance of closed-loop systems with long deadtime, the use of deadtime compensator (DTC) and model-based predictive controllers (MPCs) such as Smith Predictor and internal model control (IMC) have been proposed by researchers [1]. The drawbacks of DTCs and MPCs are that they require the exact model of the system and the number of tunable parameters is increased as compared to the PID only structure [2–4]. This adds to the design complexity.

S. M. Hassan et al., *Hybrid PID Based Predictive Control Strategies for WirelessHART Networked Control Systems*, Studies in Systems, Decision and Control 293,
https://doi.org/10.1007/978-3-030-47737-0_2

Table 2.1 Tunable parameters of selected controllers

Controller	Model parameter			Controller parameter		
PID	–	–	–	K_c	T_i	K_d
PPI	–	–	L_p	K_c	T_i	–
Smith predictor	K	T	L_p	K_c	T_i	–
IMC	K	T	L_p	–	T_{cl}	–

In order to avoid the complexity of the DTCs, MPCs and the poor performance of the PIDs, a special deadtime compensator; the PPI controller was proposed in [5]. The controller can well serve as a settlement between the complex MPC and the simple but poor performance PID as far as long deadtime and varying network delays are concerned [6]. This controller was adopted for wireless networked control systems characterized by network delay and long process deadtime in our earlier work [7]. The advantages of the PPI controller are that it allows for a model mismatch, it can handle integrating processes and has fewer tunable parameters than the other model-based compensators [6]. Furthermore, while both the Smith Predictor and the IMC controllers require a systematic experiment to identify the parameters of the process for the design of the controller, the PPI controller does not require this identification process. Hence the parameters of the controller can be manually tuned. In comparison with the PID controller, prediction in PPI controller is possible even with long deadtime without amplifying high-frequency noise. Thus the PPI is expected to give faster responses than the PID if the deadtime is long. Prediction in the PPI is achieved through low pass filtering of the control signal. Furthermore, the controller is guaranteed to improve systemâs control performance and can work well in systems with varying delay. The implementation of the controller is also cheaper as compared to the model based controllers [5].

To demonstrate the similarities of the PPI with the PID in terms of simplicity, consider for example controlling first order dead time (FODT) system commonly used to represent practical systems. Five and four parameters are required to be tuned if Smith Predictor and Internal model control are used respectively. On the other hand, three parameters are required to be tuned for both PPI and PID controllers as shown in Table 2.1. As seen from the table, while both IMC and requires full knowledge of the model parameters, the PPI only requires the estimate of the delay in the system. The other two parameters of the PPI can be tuned the same way as tuning PI controller.

High-frequency measurement noise can degenerate the performance of control system by generating control activity that may lead to wearing of the actuator [8, 9]. Other effects of the noise apart from the wearing of the actuator are heat dissipation, acoustic sound, increase production cost and a reduction in overall control precision [10]. Thus, DTCs, MPCs and PIDs can be designed to have good load regulation and robustness. However, when this high-frequency measurement noise is involved, the load regulation capability need to be backed by additional filtering. This is to curtail straining of the actuator by large signals due to undesired control activity caused by the noise. The typical PPI controller adopted for the wireless environment is not immune from this effect [7]. Thus, this work proposes an improvement to the PPI structure by incorporating a filter into the design. The filter if appropriately chosen will improve a closed loop control performance [11]. This implies the need for an additional tunable parameter (filter time constant). Despite the additional parameter, the FPPI still retains its comparative advantages over both the PID and MPCs.

Several works regarding delay compensation techniques such as IMCs, DTCs and thus PPI controller have been reported in the literature. A comprehensive survey of several DTCs can be found in [1]. In [12] an improvement to the robustness of deadtime compensating PI controller is presented. The controller incorporates a first order filter in the feedback loop of the structure. This increases the number of tunable parameters of the Smith Predictor from 5 to 6. The structure of the robust PPI controller is similar to that of traditional Smith Predictor and does require the nominal plant model for its implementation. A new PPI controller with additional filtering has been proposed in [13]. Here, the prediction is achieved through the use of disturbance observer (DO). Comparison between two degrees of freedom (2DOF) PI controller and a predictive disturbance observer (PDO) based filtered PI controller is presented in [14]. In the paper, it was shown that enhanced loop performance can be achieved with the PDO based PPI. Authors in [15] proposed adding a prediction algorithm to the traditional PID algorithm. Here, a noise filter is considered as part of the design parameter. However, only one simulation example is considered with a short deadtime of less than 1s. In [6], a comprehensive comparison between PPI controller and robust PID is presented. The authors highlighted the potential of using measurement filter for PPI controller. In [16] the PPI controller has been used alongside typical PID to simulate for control of WirelessHART network subject to clock drift. However, this work has been based on pure simulation thus, there is no reference to real-time delay.

2.3 Development of FPPI Controller For WHNCS

This section presents the application of PPI control strategy for an environment characterized by process deadtime, random network delay, and measurement noise. An attempt will be made to improve the performance of the PPI by proposing an improvement to its structure through adding a filter to the structure. The PPI controller has the advantages of being simple as the PID and also effective with both random delay and model mismatches.

2.3.1 Typical PPI Structure

Assuming the plant's transfer function is a first order dead time given as

$$G_p(s) = \frac{K}{1+Ts}e^{-sL_P}, \tag{2.1}$$

The transfer function of the closed loop of Fig. 1.8 without the wireless network is given as

$$G_o(s) = \frac{G_P(s)G_c(s)}{1+G_P(s)G_c(s)}, \tag{2.2}$$

Thus from Eq. (2.2), $G_c(s)$ is obtained as

$$G_c(s) = \frac{1}{G_p(s)}\frac{G_o(s)}{1-G_o(s)}, \tag{2.3}$$

Define the desired closed loop transfer function as

$$G_o(s) = \frac{e^{-sL_P}}{1+Ts}, \tag{2.4}$$

Using Eq. (2.4) in (2.3), the controller is expressed as

$$G_c(s) = \frac{1+sT}{K(1+Ts-e^{-sL_P})} \tag{2.5}$$

Expressing the controller $G_c(s)$ in Eq. (2.5) in terms of the relationship between its input $E(s)$ and its output $U(s)$ we have

$$(1+Ts-e^{-sL_P})U(s) = \frac{1}{K}(1+sT)E(s), \tag{2.6}$$

Thus, Eq. (2.6) can be expressed as follows:

$$U(s) = \frac{1}{K}(1+sT)E(s) - \frac{1}{sT}(1-e^{-sL_P})U(s), \tag{2.7}$$

From Eq. (2.7), it can be deduced that a PI controller acts upon the error signal $E(s)$ and the prediction is achieved via low-pass filtering of the control signal $U(s)$. Again, when the delay term $L_P = 0$, the controller is reduced to a PI controller with gain $K_C K = \alpha$ and $T_i = \beta T$. α and β are tunable parameters. For the first order system these constants could be chosen to be unity each.

For the case when $\alpha = \beta = 1$ and $L_P > 0$, Eq. (2.7) can be conveniently written as

$$U(s) = K_c E(s) + \frac{1}{1+T_i s}e^{-sL_P}U(s). \tag{2.8}$$

The implementation of Eq. (2.8) is shown in Fig. 2.1 with $F(s) = 0$.

Furthermore, Eq. (2.8) can be written in the form of a PPI controller transfer function and be factored in to $G_c(s) = C_{PI}(s)C_{pred}(s)$ as follows:

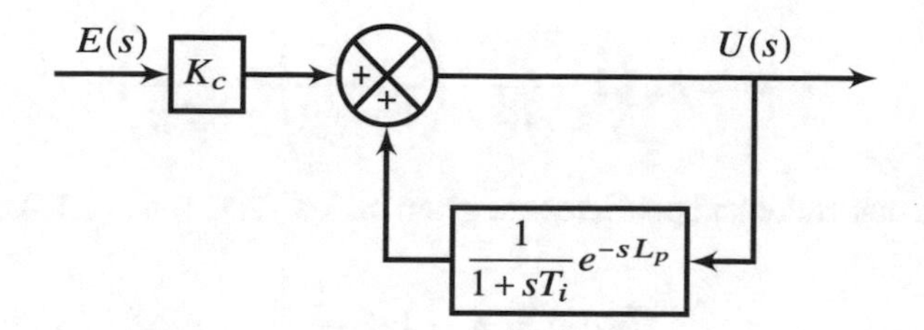

Fig. 2.1 Implementation of typical PPI controller

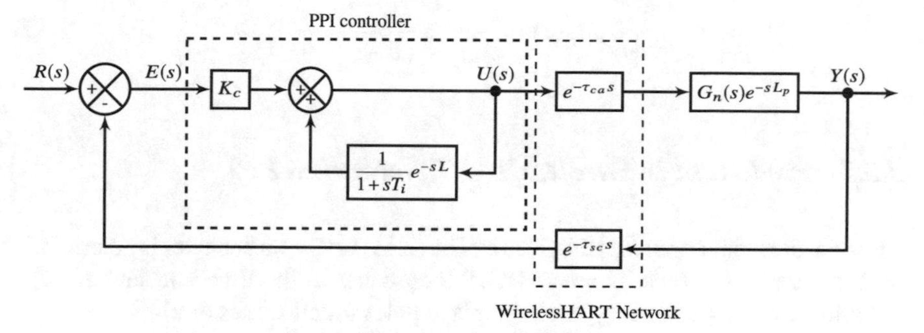

Fig. 2.2 PPI controller in WirelessHART network set-up

$$G_c(s) = K_c\left(1+\frac{1}{T_i s}\right)\left(\frac{1}{1+\frac{1}{T_i s}(1-e^{-sL_P})}\right), \tag{2.9}$$

where, $C_{PI}(s) = K_c(1+\frac{1}{T_i s})$, is the PI controller and $C_{pred}(s) = \frac{1}{1+\frac{1}{T_i s}(1-e^{-sL_P})}$ is the predictor.

2.3.2 WirelessHART Networked PPI Structure

Consider the process in Fig. 2.2 with the wireless network characterized by both network τ_N delay and process deadtime L_P, the total delay in the loop is given in Eq. (1.3).

If $G_c(s)$ a PPI controller and the delay is the total loop delay i.e. $L = \tau_N + L_P$, the PPI controller for the wireless systems can be expressed as Eq. (2.10).

$$U(s) = K_c E(s) + \frac{1}{1+Ts}e^{-sL}U(s), \tag{2.10}$$

Furthermore, Eq. (2.10) can be written in the form of a PPI controller transfer function and be factored in to $G_c(s) = C_{PI}(s)C_{pred}(s)$ as follows:

$$G_c(s) = K_c\left(1 + \frac{1}{T_i s}\right)\left(\frac{1}{1 + \frac{1}{T_i s}(1 - e^{-sL})}\right), \tag{2.11}$$

where the PI controller and predictor are given in Eqs. (2.12) and (2.13) respectively.

$$C_{PI}(s) = K_c\left(1 + \frac{1}{T_i s}\right) \tag{2.12}$$

$$C_{pred}(s) = \frac{1}{1 + \frac{1}{T_i s}(1 - e^{-sL})} \tag{2.13}$$

2.3.3 Prediction in WirelessHART Networked PPI

Consider the predictor in the PPI given by Eq. (2.13) below, its behavior is determined by the ratio of the total WirelessHART loop delay to the time constant (L/T_i). Additionally, it contains only left hand plane poles for all values of T_i.

A series expansion of Eq. (2.13) for small value of s gives

$$C_{pred}(s) \approx \frac{1}{1 + \frac{L_p}{T_i} - \frac{sT_i\left(\frac{L}{T_i}\right)^2}{2} + \cdots},$$

$$C_{pred}(s) \approx \frac{1}{1 + \frac{L_p}{T_i}}\left(1 + \frac{1}{2}\frac{\left(\frac{L}{T_i}\right)^2}{\frac{L}{T_i}} T_i s + \cdots\right), \tag{2.14}$$

$$C_{pred}(0) = \frac{1}{1 + \frac{L}{T_i}}, \tag{2.15}$$

The static gain of the predictor is given in Eq. (2.15). Furthermore, the PPI controller is only equal to PI-controller for certain parameter values where model matching is necessary. Thus the performance of the PPI can be improved even without considering model matching. The frequency plot of the predictor (see Fig. 2.3) shows that for high frequencies, the gain of the predictor approaches unity while the phase advance falls rapidly to zero for different delay times. This implies that the phase and magnitude of the predictor remain constant at 1 and 0° respectively regardless of the delay values at high frequencies. This information on the predictor will guide the selection of filter order for the PPI controller.

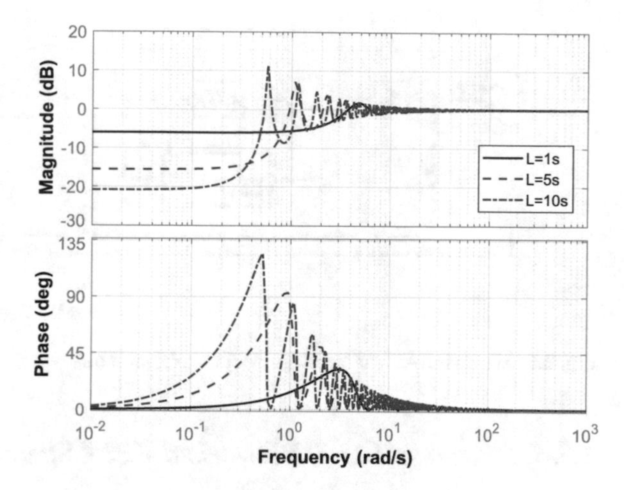

Fig. 2.3 Predictor frequency plot for $L = 1$ s, 5 s, 10 s

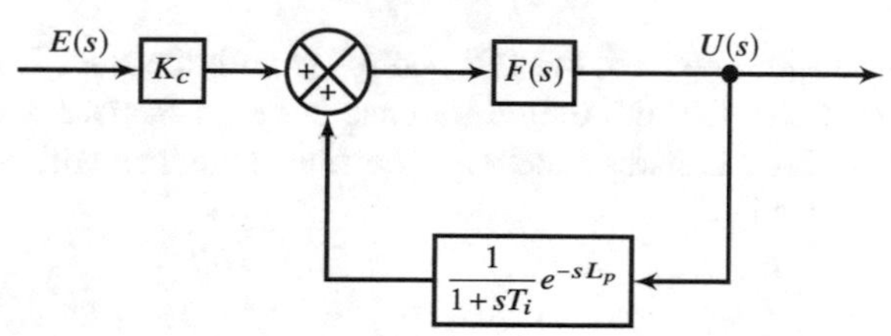

Fig. 2.4 Implementation of FPPI controller

2.3.4 *Filtered Predictive PI Structure*

The Filtered Predictive PI (FPPI) structure is given in shown in Fig. 2.4. The difference between the conventional PPI structure and the FPPI is the inclusion of the filter term $F(s)$ which will help curtail the effect of noise and oscillation induced by higher order systems and stochastic nature of the network. It should be noted that the design of PPI is based on FOPDT systems.

From the figure, the transfer function from E to U is given as

$$\frac{U(s)}{E(s)} = \frac{K_c F(s)}{1 - \frac{e^{-sL}}{1+sT_i} F(s)}, \tag{2.16}$$

where $F(s)$ is a filter transfer function. Thus, (2.16) can be expressed as

$$U(s) = \left(K_c E(s) + \frac{1}{1 + sT_i} e^{-sL} U(s) \right) F(s), \tag{2.17}$$

From Eq. (2.17), it can be seen that the PPI control action is passed through a filter $F(s)$ to achieve the control action of the FPPI controller. This implies that both the error signal and prediction term are filtered to achieve better performance.

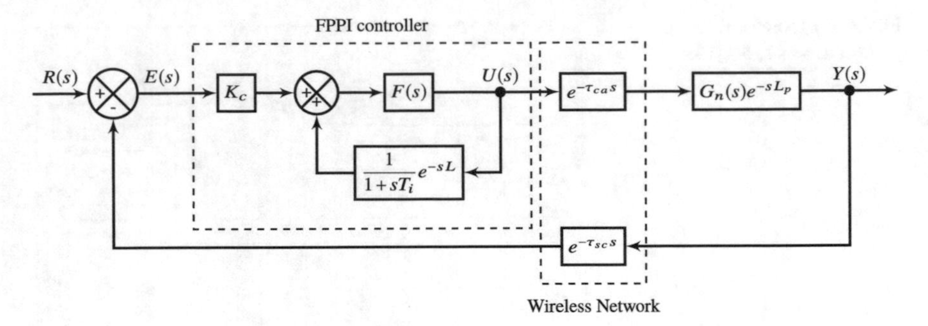

Fig. 2.5 WirelessHART networked FPPI control structure

2.3.5 WirelessHART Networked FPPI Structure

Consider Fig. 2.5, the PPI controller with the filter is given in Eq. (2.17). It should be noted that the only difference between Sects. 2.3.2 and 2.3.5 is the addition of the filter transfer function. The filter structure will be discussed in the following subsection.

2.3.6 Filter Structure

Consider the predictor in Eq. (2.13) of the PPI controller, as shown earlier, the gain of the predictor approaches unity at high frequencies. This clearly indicates that first order measurement filter should be used for the PPI controller. This point has also been suggested by [6]. Therefore, the following filter structure is used:

$$F(s) = \frac{1}{1 + sT_f}, \text{ for } T_f > 0, \tag{2.18}$$

where $T_f = \epsilon L$ and $\epsilon > 0$.

The digital implementation of the filter can be achieved by using the following recursive relationships:

$$y(1) = u(1), \tag{2.19}$$

for $k > 1$,

$$y(k) = (1 - \gamma)y(k-1) + \gamma u(k), \tag{2.20}$$

where $\gamma = \frac{h}{T_f + h}$ is the filter constant, h is the sampling period and should be chosen to be $h \leq \frac{T_f}{5}$ (see [17]).

The above filter is also referred to as exponentially weighted moving average filter. In this kind of filter, the weighting factor γ decreases exponentially as the time progresses.

2.3.7 Robustness and Stability Analysis

For the robustness analysis of the proposed approach, the extended sensitivity and its complementary functions method as proposed by [18] are considered. In this method, which has also been adopted by same authors in [6], the model uncertainties were divided into deadtime and non-deadtime based. The robustness computation is established on the open loop transfer function based on Nyquist stability criterion [19].

If Eq. (2.17) is the control action of the FPPI and is denoted now as the controller $G_c(s)$, the total delay in Fig. 2.5 is given as Eq. (1.3). Under nominal conditions, the entire process model including network delays can be expressed as

$$G(s) = G_n(s)e^{-sL}, \tag{2.21}$$

where, $G_n(s)$ is the delay free process.

Consider some deviations from a nominal condition where there is variation in both process deadtime and network-induced delays. Assuming that the delay error is $\Delta L \subset [\Delta L_{min}, \Delta L_{max}]$. Assume also that the multiplicative uncertainty between the nominal process $G_n(s)$ and the real process $G(s)$ is $\Delta G(s)$. Furthermore, it is assumed that the process model together with uncertainties is norm bounded and can be written as

$$G(s) = G_n(s)\left(1 + \frac{\Delta G(s)}{G_n(s)}\right)e^{-s(L+\Delta L)}, \tag{2.22}$$

If the controller of the system is considered to be $G_c(s)$, the nominal open loop in the frequency domain given as $G_c(i\omega)G(i\omega)$ is thus assumed to be stable and also norm bounded.

Consider the Nyquist diagram of the nominal open system (G_cG) shown in Fig. 2.6 with uncertainty in the delay ΔL, if point A is rotated through angle $-\omega\Delta L$ and moved slightly to any direction $|G_c\Delta G(i\omega)| = |G_c\Delta G(i\omega)e^{i\omega(L+\Delta L)}|$, it will stay within a circle defined by center $G_cG(i\omega)e^{i\omega(\Delta L)}$ and radius $|G_c\Delta G(i\omega)|_\infty$.

The distance from center $G_cG(i\omega)e^{i\omega(\Delta L)}$ to the critical point -1 is $|1 + G_c\Delta G(i\omega)e^{i\omega(\Delta L)}|$. This indicates that the upset $G_c\Delta G(i\omega)e^{i\omega(L+\Delta L)}$ will not drive the system unstable as long as

$$|G_c\Delta G(i\omega)| < |1 + G_cG(i\omega)e^{i\omega(\Delta L)}|, \quad \forall\ \omega, \Delta G, \Delta L \tag{2.23}$$

Dividing Eq. (2.23) by G_cG_n and assuming $e^{-i\omega(L+\Delta L)} = 1$, the equation can be written as

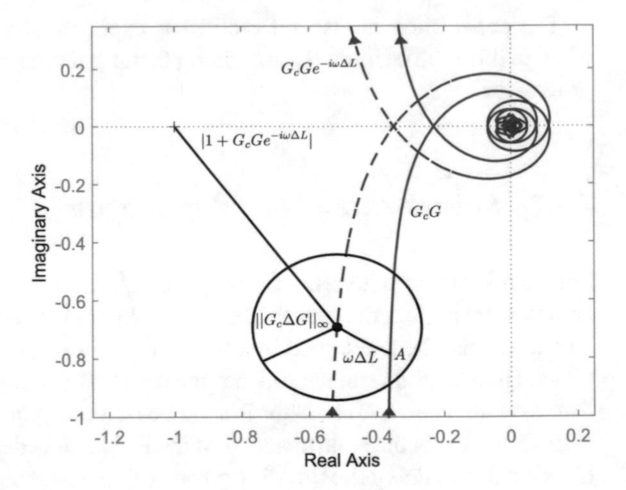

Fig. 2.6 Open loop transfer function Nyquist plot for nominal system and its uncertainty due to respective variation in process ΔG and total network delay ΔL

$$\left|\frac{1+Gc(i\omega)G(i\omega)e^{-i\omega(\Delta L)}}{Gc(i\omega)G(i\omega)e^{-i\omega(\Delta L)}}\right| > \left|\frac{\Delta G(i\omega)}{G_n(i\omega)}\right|, \tag{2.24}$$

Defining the extended complementary sensitivity function as the inverse of left hand side of Eq. (2.24) we have

$$T(s, \Delta L) = \frac{G_c(s)G(s)e^{-s\Delta L}}{1+G_c(s)G(s)e^{-s\Delta L}}, \tag{2.25}$$

Therefore, the condition for robust stability can be given as

$$\left\|\frac{\Delta G(s)}{G_n(s)}T(s, \Delta L)\right\|_\infty < 1, \ \Delta L \in [\Delta L_{min}, \Delta L_{max}]. \tag{2.26}$$

In the same way, if the inverse multiplicative uncertainty is considered, the process model can as well be written as

$$G(s) = G_n(s)\left(1+\frac{\Delta G(s)}{G_n(s)}\right)^{-1} e^{-s(L+\Delta L)}, \tag{2.27}$$

Therefore, robust stability condition based on inverse multiplicative uncertainty can be given as

$$\left|1+Gc(i\omega)G_n(i\omega)\left(1+\frac{\Delta G(i\omega)}{G_n(i\omega)}\right)^{-1} e^{-i\omega(L+\Delta L)}\right| > 0$$

the above inequality can be written as

$$\left| 1 + \frac{\Delta G(i\omega)}{G_n(i\omega)} + Gc(i\omega)G_n(i\omega)e^{-i\omega(L+\Delta L)} \right| > 0, \tag{2.28}$$

$\forall\ \Delta L \in [\Delta L_{min}, \Delta L_{max}]$, $\Delta G(i\omega)$ and ω.

Equation (2.28) can be expressed in the following form since $\Delta G(i\omega)$ can take any direction in the complex plane:

$$\left| 1 + Gc(i\omega)G_n(i\omega)e^{-i\omega(L+\Delta L)} \right| - \left| \frac{\Delta G(i\omega)}{G_n(i\omega)} \right| > 0, \tag{2.29}$$

The extended sensitivity function is thus defined as the inverse of the first term in Eq. (2.29) as follows

$$S(s, \Delta L) = \frac{1}{1 + G_c(s)G(s)e^{-s\Delta L}}, \tag{2.30}$$

Thus, the robust stability condition derived from both Eqs. (2.29) and (2.30) is

$$\left\| \frac{\Delta G(s)}{G_n(s)} S(s, \Delta L) \right\|_\infty < 1,\ \Delta L \in [\Delta L_{min}, \Delta L_{max}], \tag{2.31}$$

It should be noted that the controller $G_c(s) = C_{PI}(s)C_{Pred}(s)F(s)$ hence inclusion of the filter function which is essentially part of the open loop that is stable does not cause instability of the system.

2.3.8 Tuning of Controller Gain and Selection of Filter Time Constant

In the proposed FPPI, it is assumed that the parameters of the controller and those of the process are related by the following equations:

$$K_c K = \alpha \tag{2.32}$$

$$T_i = \beta T \tag{2.33}$$

where design parameters α and β are chosen as unity each [5].

Although this condition suffices for first order systems, it is however, a limitation while considering higher order systems. Thus, $\alpha \neq 1$ and $\beta \neq 1$. To obtain better setpoint tracking of the controller, α and β need to be tuned especially for higher order systems. The filter time constant T_f is another parameter obtained through trial and error by using the relationship $T_f = \epsilon L$. However, this criterion only suggests that $\epsilon > 0$. Thus, the need to appropriately select the filter time constant through tuning.

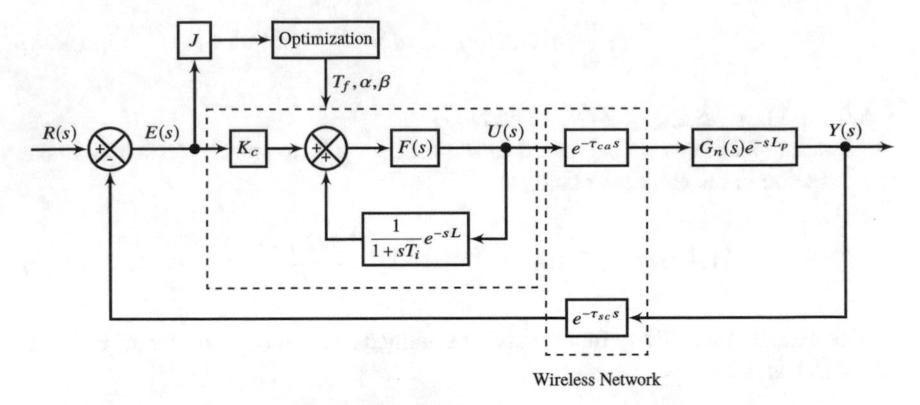

Fig. 2.7 Optimal wireless networked FPPI control structure

In order to obtain an optimal FPPI controller, the proposed HABF-APSO algorithm given in Chap. 5 will be employed to tune the values of α, β and T_f. The structure of Fig. 2.7 will be used for the tuning.

2.4 Simulation Results for FPPI Controller

In this section, two simulation approaches namely pure simulation and hardware-in-the-loop simulation will be considered. Results of the proposed FPPI approach will be compared with those of Smith predictor, PPI and PI controllers. Notably, the robustness of the proposed FPPI controller to parametric modeling error for the first order model will be considered. In this analysis, commonly encountered process models given in (2.34)–(2.37) that represent the behaviour and dynamics of practical plants in the industry are considered for simulation. Similar models can be found in [4, 5, 20]. In each case, time domain analysis of the performance of the controller is presented.

$$G_1(s) = \frac{1}{2s+1}e^{-4s}, \tag{2.34}$$

$$G_2(s) = \frac{1}{(s+1)^2}e^{-4s}, \tag{2.35}$$

$$G_3(s) = \frac{1}{(s+1)^3}e^{-5s}, \tag{2.36}$$

$$G_4(s) = \frac{1}{(1+s)(1+0.5s)(1+0.25s)(1+0.125s)}e^{-5s}. \tag{2.37}$$

Before the simulation results, the experimental set up and procedure for the network delay measurement will be discussed. The delay obtained from the measurement will be subsequently used for the pure simulation.

2.4.1 Network Delay Measurement Procedures and Experimental Set-Up

This section explains briefly the procedures to measure the WirelessHART network delay and describes the experimental setup and simulations with the WirelessHART Hardware in the loop simulator. The section also describes briefly the architecture of the simulator. Two WirelessHART network development and evaluation kits developed by RF Monolithics (RFM) and Linear Technology are used for experimentation in this work. The RFM kit was used to ascertain the level of network-induced delay in the system while the Linear Technology kit was used alongside Hybrid simulator to verify the effectiveness of the proposed method to real-time variation in delay.

The RFM WirelessHART development kit is used to measure the network induced delays. The experimental set-up consists of a network manager (Implemented in a computer), XG2510HE gateway and an XDM2510H field node otherwise known as mote as shown in Fig. 2.9. The schematics for the experimental set-up is also shown in Fig. 2.10. In both figures, the host computer is connected with the gateway through the RJ-45 cable while the gateway communicates with the mote wirelessly, is the delay from the gateway to the mote, while is the delay from mote to the gateway. The two delays can be obtained in the gateway by sending the command "*exec getLatency MACaddress*" where "*MACaddress*" is the MAC address of the node in the gateway. These delays have been measured by getting the difference in timestamp between two consecutive measurements see [21] (Fig. 2.8).

To obtain the delay information from the gateway in MATLAB, the Secure Shell (SSH2) software has been used as an interface between the MATLAB and the gateway. In this way a secured communication is established between the host application, MATLAB, and the gateway. To avoid contention between MATLAB and the gateway, the update rate between the gateway and MATLAB for the information was set to three seconds. To read the information stored in the gateway in MATLAB environment, the SSH2 command ssh2_config ('192.168.99.100', '*dustcli*', '*dustcli*') is used. In this command, the IP address of the gateway is the first argument of the command while the username and login password are the second and third arguments respectively. The experiment procedure flowchart is shown in diagram of Fig. 2.8.

Here, emphasis on the ability of the proposed FPPI controller to suppress the effect of noise on the PPI will be given. All systems are simulated to a unit step input signal with a disturbance of magnitude 0.5 is applied at 200 s. At the output, a zero mean white noise signal of small magnitude 0.0001 is added at the output to simulate for possible noise scenario. Furthermore, the proposed FPPI controller is tested for its ability to track changing setpoint for all systems, while its robustness to parametric model mismatch is tested for the first-order system.

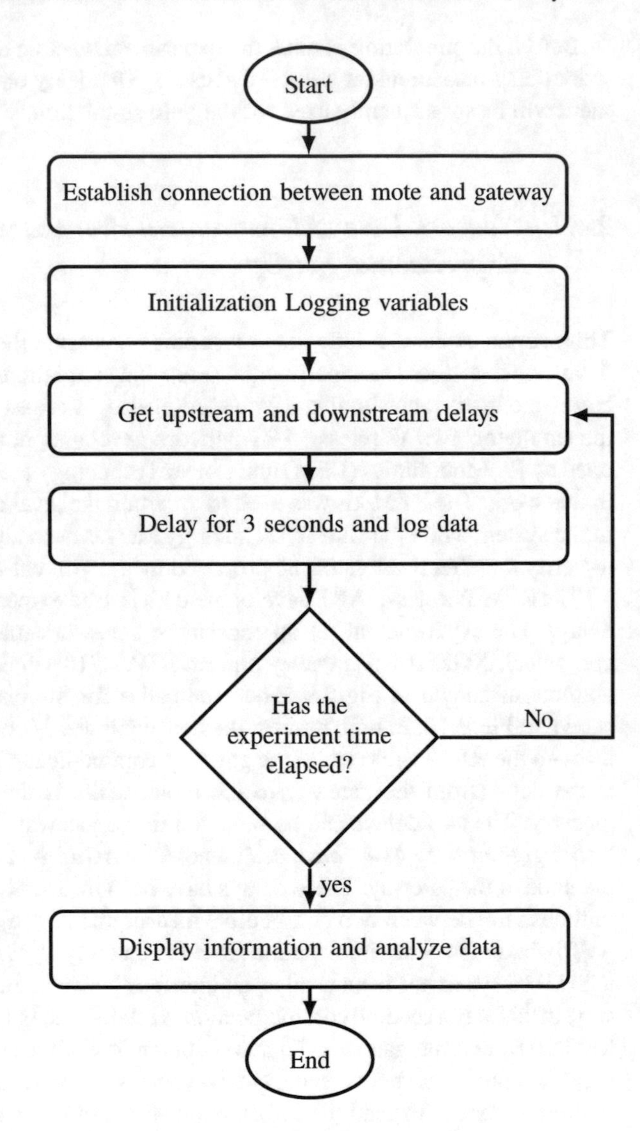

Fig. 2.8 Delay measurement procedure

2.4.2 *Controller Parameters*

The parameters of the PPI, PI and Smith Predictor controllers are given in Table 2.2 while the tuned parameters of the FPPI are presented in Table 2.3. For the FPPI controller, the parameters are tuned using the HABF-APSO algorithm discussed in Chap. 4. The parameters were obtained from the plant models.

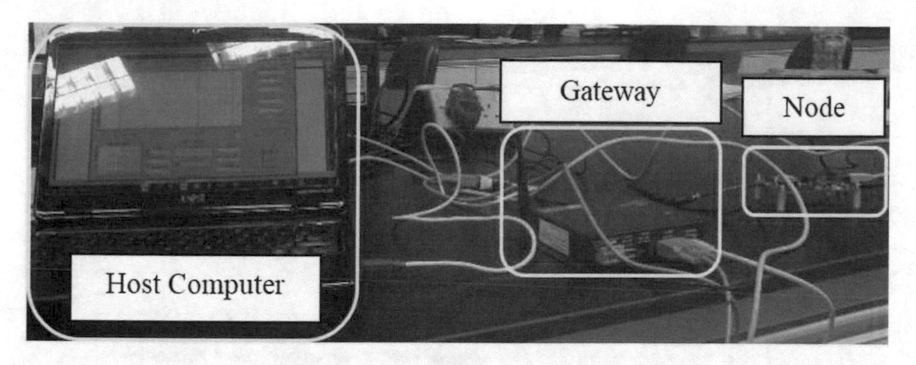

Fig. 2.9 Experimental set-up to measure network induce delay in WirelessHART controlled network

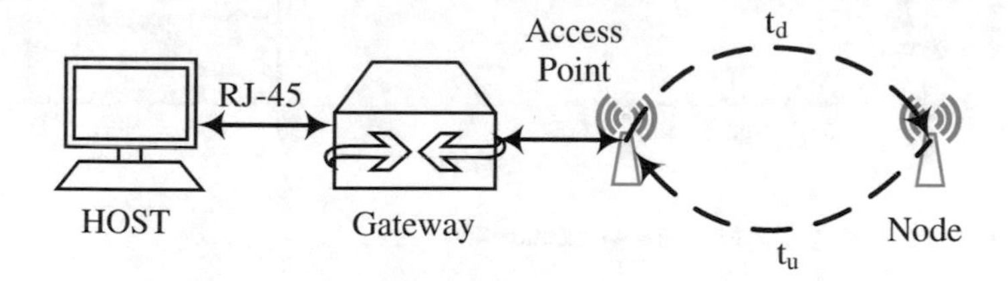

Fig. 2.10 Schematics for WirelessHART network induced delay measurement

Table 2.2 PPI, PI and Smith predictor parameters

Parameter	Plant			
	$G_1(s)$	$G_2(s)$	$G_3(s)$	$G_4(s)$
K_c	0.125	1	1	1
T_i	9.31	1.3	2	1.5
L_p	10	5	5	5
K_p	0.0365	0.271	0.291	0.280
K_i	0.0047	0.0777	0.0721	0.0793

Table 2.3 Tuned FPPI controller parameters

Parameter	Plant			
	$G_1(s)$	$G_2(s)$	$G_3(s)$	$G_4(s)$
T_f	0.8256	2.3000	3.7586	4.6258
α	0.9235	0.8205	0.8500	0.9706
β	0.9750	0.9986	1.0059	1.2114

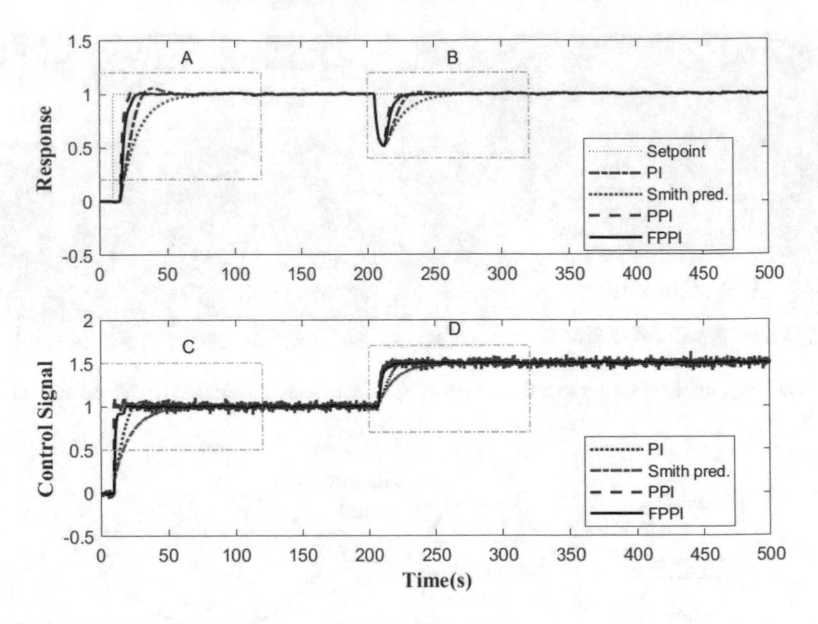

Fig. 2.11 First order system response with controllers

2.4.3 *First Order System*

The performance of the plant based on the four controllers compared is shown in Fig. 2.11 and the numerical results displayed in Table 2.4. Observing the response signals of Fig. 2.11 and numerical results of Table 2.4, it can be seen that the response of proposed FPPI has improved on both PPI and PI controllers in terms of Settling time and overshoot. Although the proposed FPPI is a bit slower in rise time than PPI due to filter effect, it is still faster than Smith Predictor and PI. Observing the control signal of the various controllers, it can be seen that the signal of the proposed FPPI is smoother than that of the PPI. On the recovery from the effect of disturbance, the proposed approach recovered in a similar way to PPI while being faster than both PI and Smith Predictor controllers. This indicates that the FPPI controller maintains the characteristics of the PPI while showing robustness to measurement noise and disturbance. These results are further revealed in the zoomed in plots of the regions of interests A, B, C and D in Fig. 2.12. In terms of the IAE, the PPI has the least error of 181.8273 followed closely by FPPI with 192.6705, then Smith Predictor with 265.5232 and lastly PI with 436.0825. It should be noted that, despite the least overshoot of Smith predictor, it is the slowest of all in terms of both rise time and recovery from disturbance.

To evaluate the tracking ability of the proposed FPPI controller to changing set-point, the plant is simulated to a changing reference signal and the result is shown in Fig. 2.13. From this figure, it can be observed that the response of the FPPI and PPI have similar tracking ability that is faster than both Smith Predictor and PI. Fur-

Table 2.4 Control performance of first order system

Controller	System's performance			
	Rise time (s)	Settling time (s)	Overshoot (%)	IAE
FPPI	7.3209	26.4117	0.7516	192.6705
PPI	4.4573	491.6822	2.3930	181.8273
Smith predictor	28.3052	65.7192	0.4993	265.5232
PI	12.2081	50.1304	5.6038	436.0825

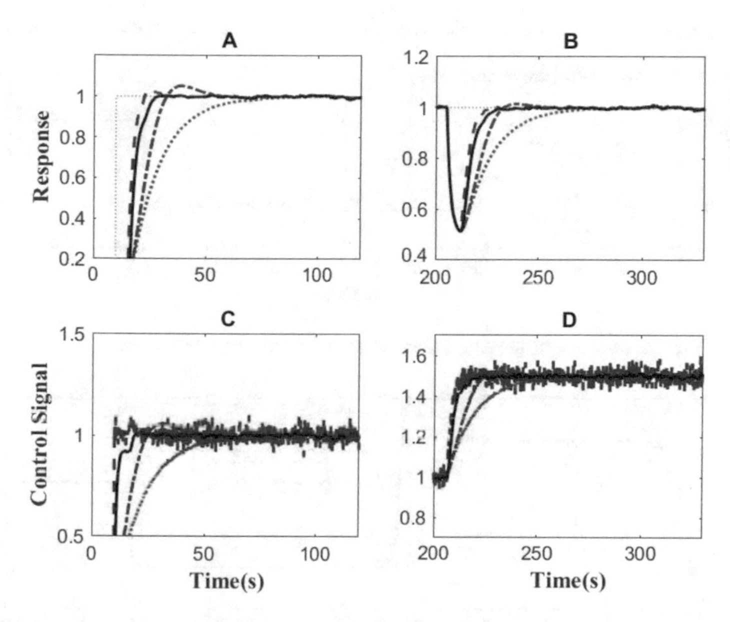

Fig. 2.12 Zoomed-in view of Fig. 2.11 for regions A, B, C and D

thermore, the control signal of the FPPI is smooth when compared to that of PPI. This is because of the filter used in the FPPI. As compared to the FPPI and PPI, the response of the PI is slow with some overshoots and undershoots while the response of the Smith Predictor is even slower but with no overshoot or undershoot.

To analyze the sensitivity of the proposed approach to model mismatch, parametric modeling error due to the mismatch in the model parameters (in this case model gain K and time constant T) is considered. The plant is simulated to both 5% increase and decrease in both model gain and time constants. The result with this perturbations is compared to that of the nominal plant and presented in Fig. 2.14. It can be observed that despite the perturbation in the model gain and time constant, the proposed controller still retained its characteristics of speed and good tracking ability.

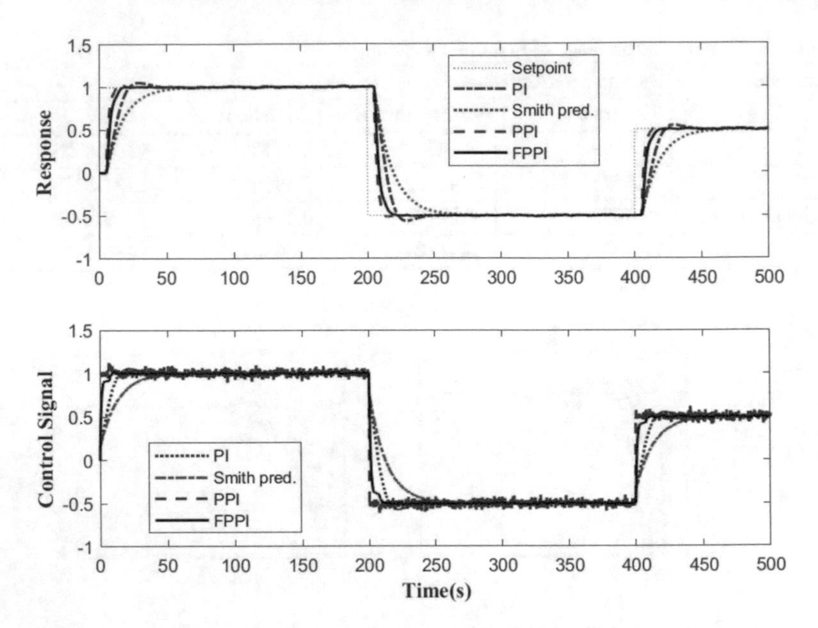

Fig. 2.13 Setpoint tracking of the first order system with various controllers

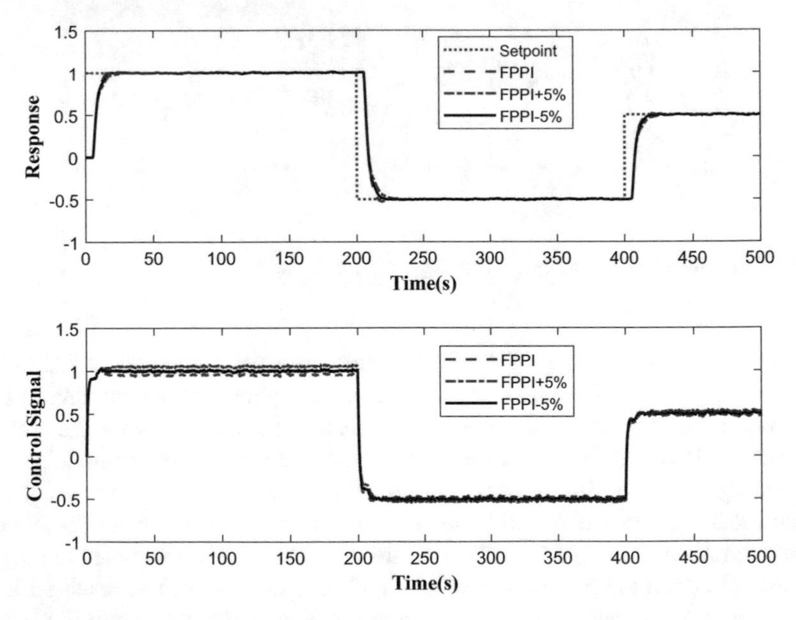

Fig. 2.14 Sensitivity of the FPPI controller to parametric modeling error (5% increase and decrease) in gain and time constants of first order system

Table 2.5 Control performance of second order system

Controller	System's performance			
	Rise time (s)	Settling time (s)	Overshoot (%)	IAE
FPPI	10.9099	33.1492	1.1933	243.6242
PPI	3.3380	45.8322	19.0248	218.1560
Smith predictor	29.9758	72.4615	0.7300	430.2872
PI	7.8350	51.4289	14.7370	270.8047

2.4.4 Second Order System

In the same way to the first order system, parameters of the various controllers are shown in Table 2.2 while the tuned parameters of the FPPI are shown in Table 2.3. Performance of the plant with the compared controllers is shown in Fig. 2.15. In the figure, regions of interest A, B, C, and D are zoomed and further highlighted in Fig. 2.16. Numerical results of the performance is shown in Table 2.5. It can be seen by observing the regions of interest that, while the PI and Smith predictors produce noisy control signals, that of the PPI is the most affected. On the other hand, the signal of the proposed approach is smoother than the other controllers compared. As expected, the response to the proposed FPPI approach is smoother with an overshoot of just around 1% as compared to those of PPI and PI with around 19 and 14% respectively.

The rise times of the four controllers are 3.33, 7.84, 29.98 and 10.41 s for the PPI, PI, Smith Predictor and FPPI controllers respectively. The longer rise time of the FPPI compared to both PPI and PI is due to the effect of filtering. Nonetheless, the proposed approach is still faster in response than the Smith Predictor although overshoot of later the is a bit smaller. Furthermore, the FPPI settles faster than the other three controllers with selling time of just above 33 s as against around 45, 51 and 71 s of PPI, PI and Smith Predictor respectively. The recovery from disturbance effect of the FPPI controller is faster than that of the Smith predictor. This recovery is also without overshoot compared to the overshoots produce by the PPI and PI as seen more clearly in Fig. 2.16.

As in the case of first order system, the tracking ability of the proposed FPPI controller to changing setpoint for this plant is analyzed by observing the result shown in Fig. 2.17. From the figure, it can be observed that the tracking ability of the FPPI from its response outperformed the other controllers compared. The response is faster than that of Smith Predictor and does not have overshoot and undershoot like the PPI and PI. The control signals of the FPPI and PI are smooth when compared to the Smith Predictor and PPI. This is because of the filter used in the FPPI while the PI compared to PID does not amplify measurement noise. However, the PI produced a response with some overshoot and undershoots as compared to the FPPI. This is visible in both response and control signals of Fig. 2.17.

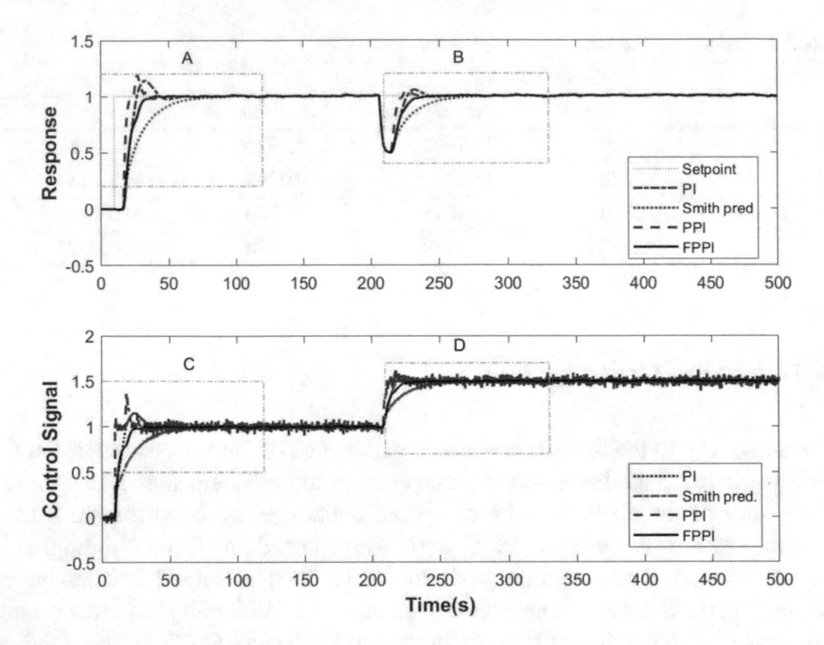

Fig. 2.15 Second order system response to FPPI, PPI, and PI controllers

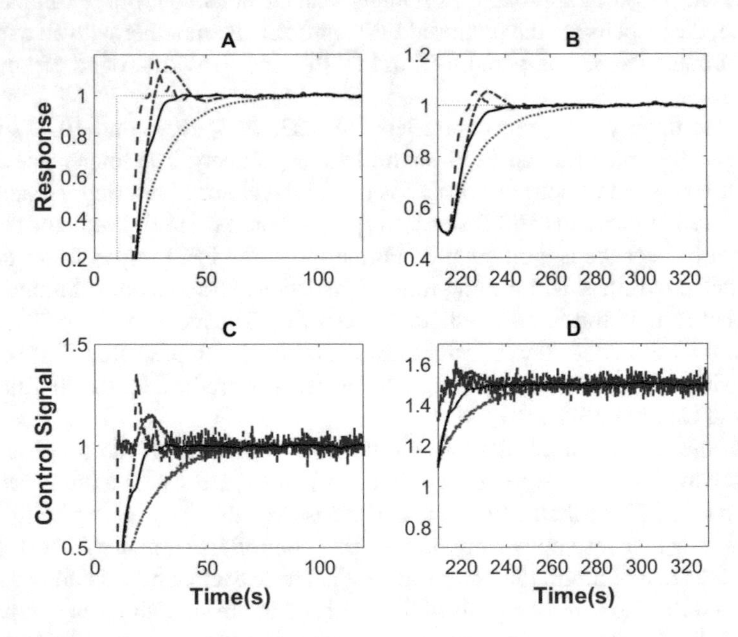

Fig. 2.16 Zoomed-in view of Fig. 2.15 for regions A, B, C and D

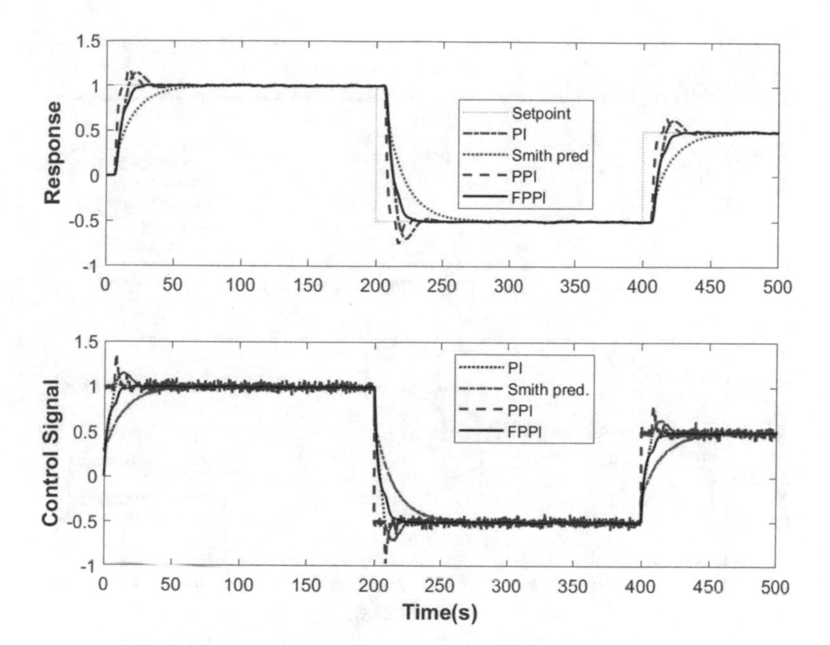

Fig. 2.17 Setpoint tracking of the second order system with various controllers

2.4.5 *Third Order System*

Similar to the case of first and second order systems, parameters of the compared controllers for the third order system are shown in Tables 2.2 and 2.3. Performance of the plant subjected to the actions of four controllers is shown in Fig. 2.18. In the figure, regions of interest A, B, C, and D are zoomed and further highlighted in Fig. 2.19. The numerical results of the performance are shown in Table 2.6.

The same observation is made here as the second order system regarding the regions of interest. Just as observed in second order system, both the PI and Smith predictors produce noisy control signals, however, the signal of the PPI is the most affected the noise. Furthermore, the response with proposed approach is smoother with an overshoot of just around 1% that is very close to the 0.7% of Smith predictor. However, this is very good as compared to those of PPI and PI with around 23 and 6.7% overshoots respectively. The response of the four controllers in increasing order are 4.15, 9.81, 12.40 and 30.78 s for the PPI, PI, FPPI and Smith Predictor controllers respectively. It should be noted that least undershoot of 0.29% is recorded for the proposed FPPI controller compared to 0.30, 0.33, and 0.78% of the Smith predictor, PI, and PPI controllers respectively. Just as observed in the second order system, the slow response of the proposed approach is due to the filter effect. Still, the response with the approach settled at 36 s which is faster than respective settling times of PPI, PI and Smith Predictor which stands each at 49, 44 and 73 s.

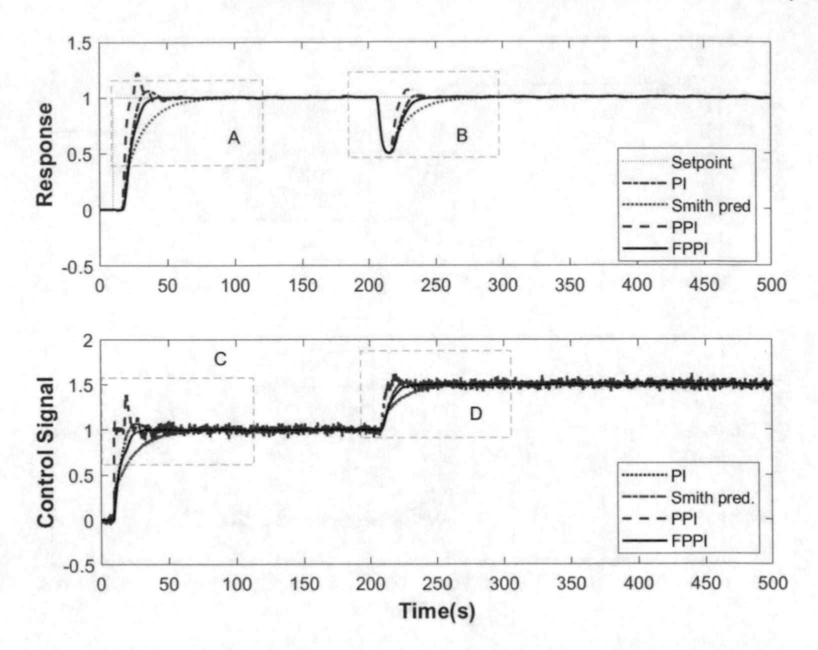

Fig. 2.18 Third order system response with various controllers

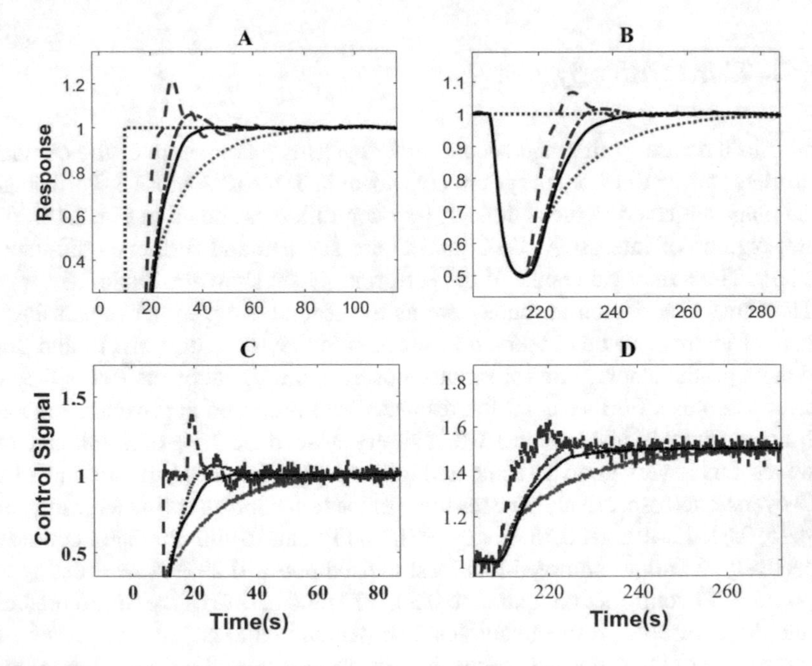

Fig. 2.19 Zoomed-in view of Fig. 2.18 for regions A, B, C and D

Table 2.6 Control performance of third order system

Controller	System's performance			
	Rise time (s)	Settling time (s)	Overshoot (%)	IAE
FPPI	12.4007	36.6080	1.1400	277.86
PPI	4.1539	49.0488	22.8517	247.07
Smith predictor	30.7813	73.4430	0.7405	443.94
PI	9.8081	44.1103	6.7137	270.49

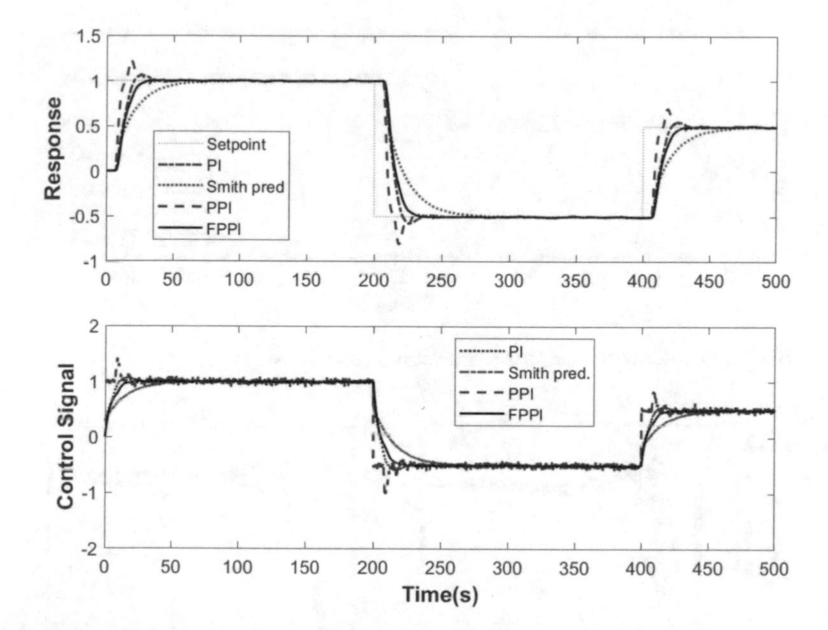

Fig. 2.20 Setpoint tracking of the third order system with various controllers

The changing setpoint tracking performance of the FPPI controller with this plant also follows similar trend the second order system. Similar observations are made with regard to the speed of response and smoothness of the control signal. Furthermore, the FPPI also recorded no overshoot or undershoot for this plant. This can be noticed when both the response and input plots in Fig. 2.20 are observed.

2.4.6 *Fourth Order System*

In a similar fashion to the first, second and third order systems, simulation result for fourth order system are presented in Fig. 2.21 while the regions of interest A, B, C, and D are zoomed in Fig. 2.22. The numerical results for the figures are presented in Table 2.7.

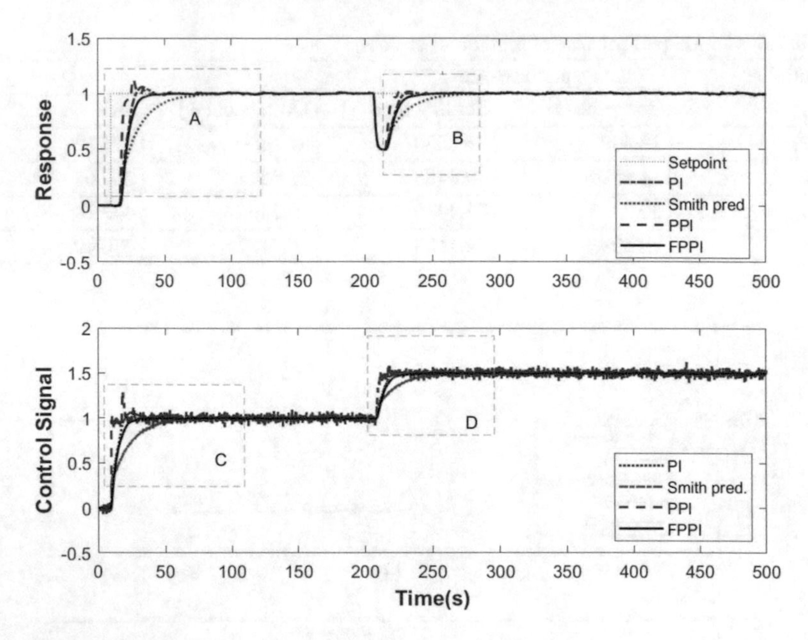

Fig. 2.21 Fourth order system response with various controllers

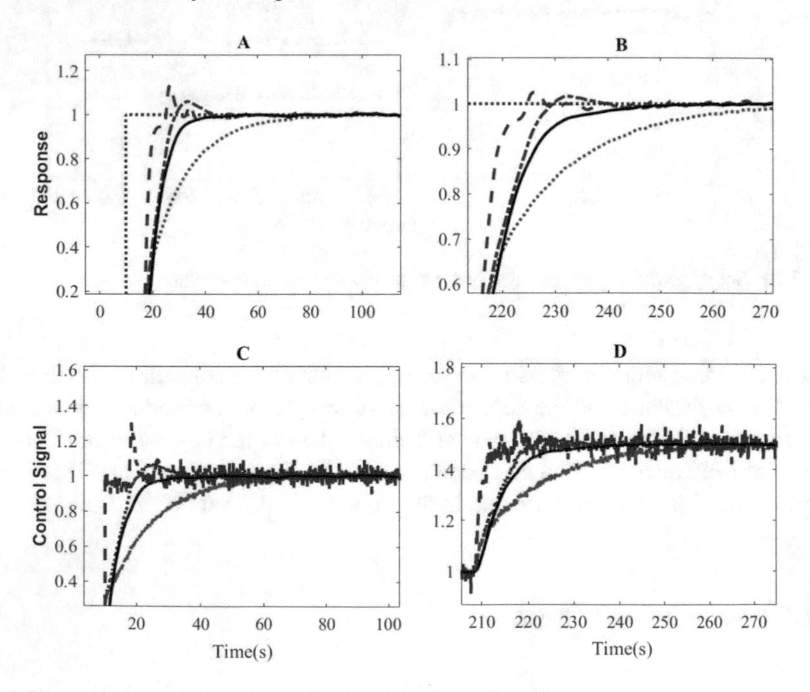

Fig. 2.22 Zoomed-in view of Fig. 2.21 for regions A, B, C and D

Table 2.7 Control performance of fourth order system

Controller	System's performance			
	Rise time (s)	Settling time (s)	Overshoot (%)	IAE
FPPI	10.6549	36.1960	1.2176	314.66853
PPI	3.2987	116.1724	13.2481	265.1365
Smith Pred.	29.4414	72.5349	0.6859	485.3059
PI	8.5599	40.4391	6.5075	299.7992

A similar pattern of results with second and third order systems are observed for this system. Again, the effect of noise on the PPI is evident from the oscillatory nature of both its control signal and response as shown in Fig. 2.22. Furthermore, the settling time of the proposed FPPI approach is shorter as compared to the other controllers. In terms of overshoot, the lowest value of 0.7% is recorded for the Smith Predictor followed by the proposed method with 1.22%. The overshoots for PPI and PI are 6.5, 13.2% respectively. It is evident from the figures that, despite the lower overshoot Smith predictor, it is however too sluggish with a rise time of 29 s. Thus, the proposed FPPI is better considering its shorter settling time and moderate rise time.

The result of the tracking ability of the proposed FPPI approach shown in Fig. 2.23 also followed a similar pattern to the results with lower order systems. Its faster than the Smith predictor. In comparison to both PI and PPI, it produced a smoother control signal and response without overshoot.

2.4.7 *Real-Time Simulation of FPPI Controller with WH-HILS*

As explained earlier, two approaches are employed to validate the controller design in the simulation environment. The first approach is the pure simulation discussed above. This method uses delay information obtained from an experiment for both controller design and implementation. On the other hand, the second approach uses delay information obtained from the previous experiment for controller design while for implementation, real-time delay information is used. The WirelessHART Hardware in the Loop Simulation (WH-HILS) is explained in detail subsequently.

The WH-HILS scheme, just like many other hardware-in-the loop simulator, allows for diagnostics as well as new control strategies to be tested before being deployed in the actual plant. This will save cost and ensure non interference of the plants' operation. Figure 2.24 shows the block diagram of the WH-HILS. Hardware-in-the-loop simulation approach has been existing for decades. For some of the recent applications of this technique see [22–25]. As compared to the traditional pure simulation, it has the advantage of using real hardware for simulation. By using real

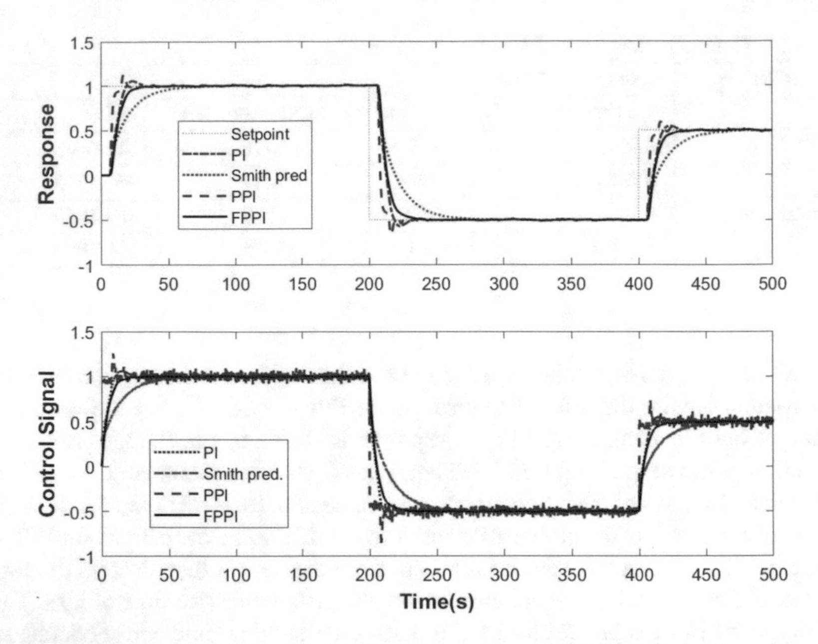

Fig. 2.23 Setpoint tracking of the fourth order system with various FPPI, PPI, Smith predictor and PI controllers

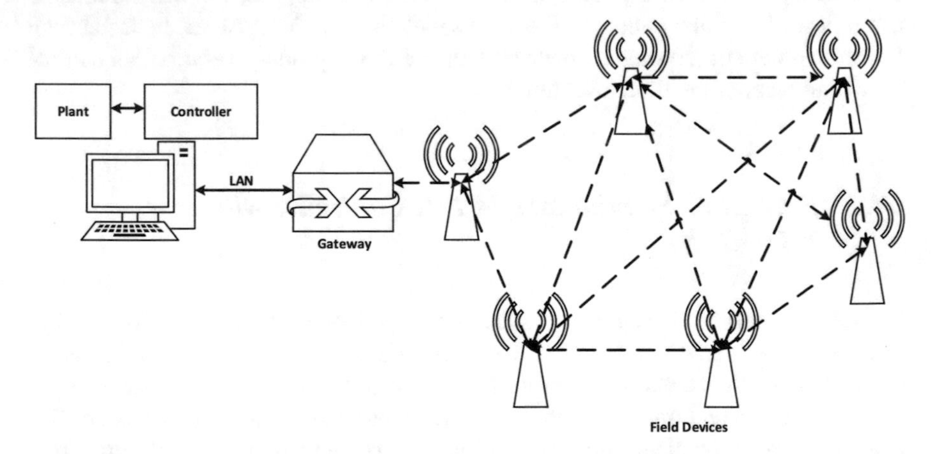

Fig. 2.24 Schematics of the WirelessHART hardware in the loop simulator

devices, model inaccuracies can be minimized, thus simulation results will be more realistic. In this work, the WH-HILS is used for simulation of WirelessHART Network Control System (WHNCS). The simulator consists of a computer, a gateway (LTP5903CEN-WHR) and several wireless nodes (DC9003-A) as shown in Fig. 2.24. As seen from the figure, the gateway is connected to the computer running MATLAB software using LAN interface as it has the highest speed among the offered communication protocols by the gateway.

The software is used to simulate virtual process plants given real-time network induced delays from the gateway. The gateway is able to measure exact network delays in communication with any mote in the network. Since delays are measured in real-time, the simulation of the plant can be run in real-time using Real-time Sync block available in Simulink. Here, it should be noted that the mesh network formed by the gateway and its motes are similar to the one used in industry as both the gateway and motes are WirelessHART certified.

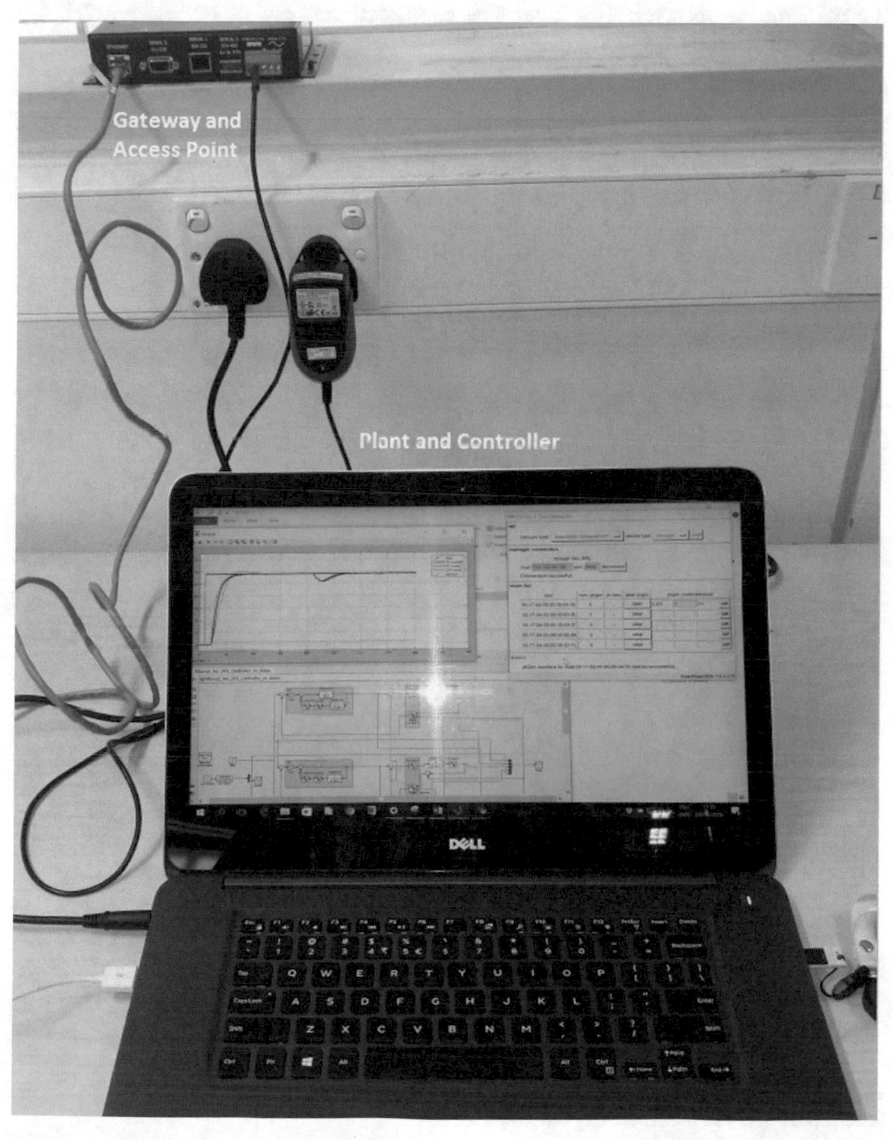

Fig. 2.25 Experimental set-up of WH-HILS: gateway/access point and plant/controller simulated in Matlab

The experimental set up for the WH-HILS process is shown in Figs. 2.25 and 2.26, each of the five motes(nodes) is placed on a Pilot process plant as shown. The real-time network-induced delay of mote with mac address '00170D000030045B' is used to simulate the plant model in Matlab. By using Real-time Sync block in Simulink, the simulation is synchronized with the computer's clock, thus it can be run in real-time. To interface the gateway with the MATLAB, a Python program is used. MATLAB R2015 and later versions natively supported this interface. The communication stack showing the interface of the MATLAB and gateway through the computer is shown in Fig. 2.27. The supported versions of Python are 2.7, 3.3, and 3.4.

Fig. 2.26 Experimental set-up of WH-HILS: location of the network motes(1, 2, 3, 4 and 5) and access point (AP) in the laboratory environment

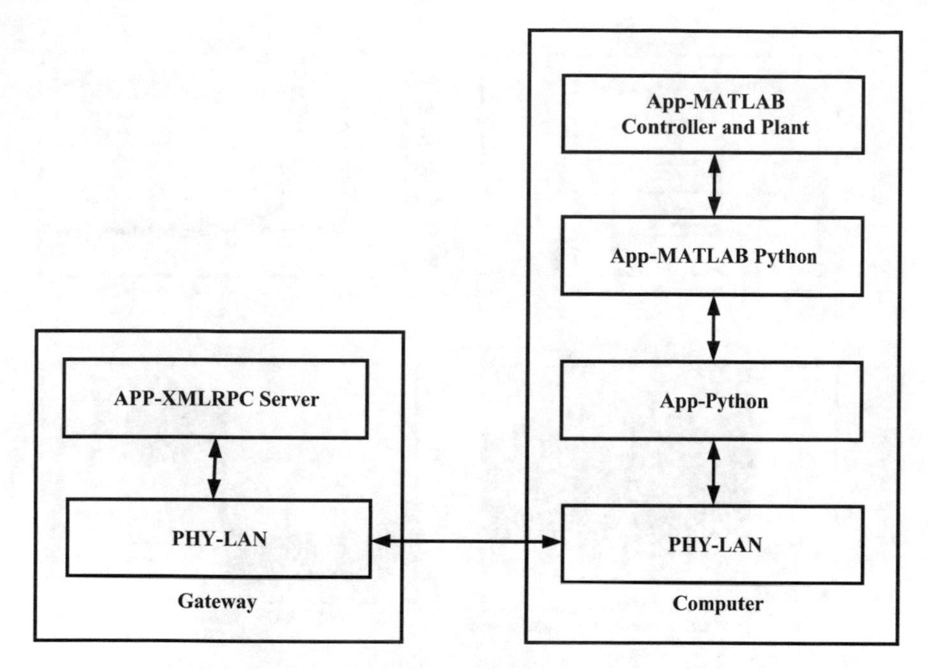

Fig. 2.27 Hardware in the loop simulators gateway-computer communication stack

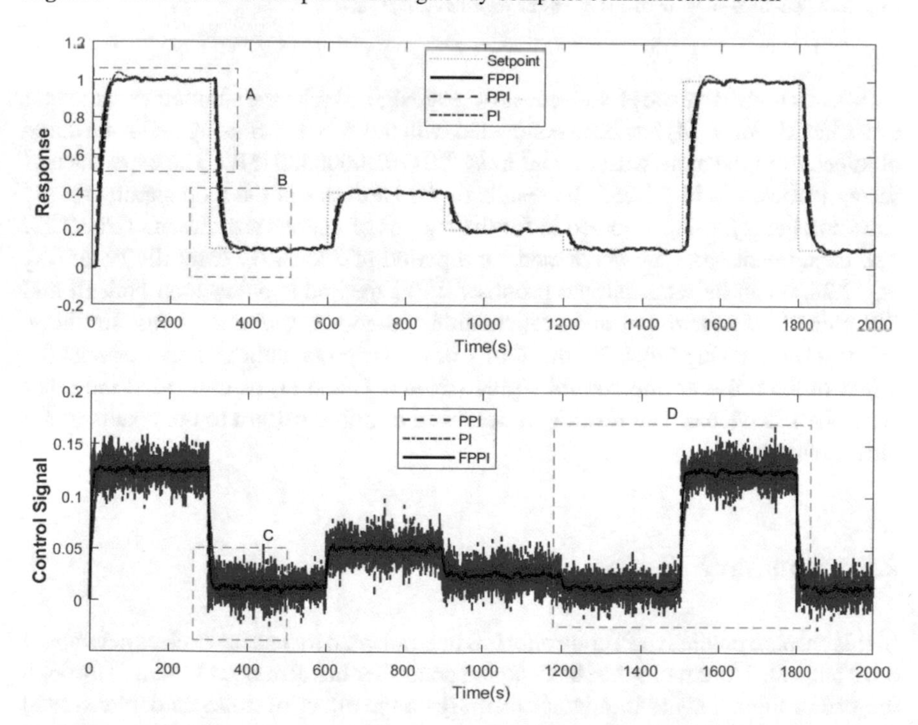

Fig. 2.28 Response of the first order system with various controllers using the WH-HILS

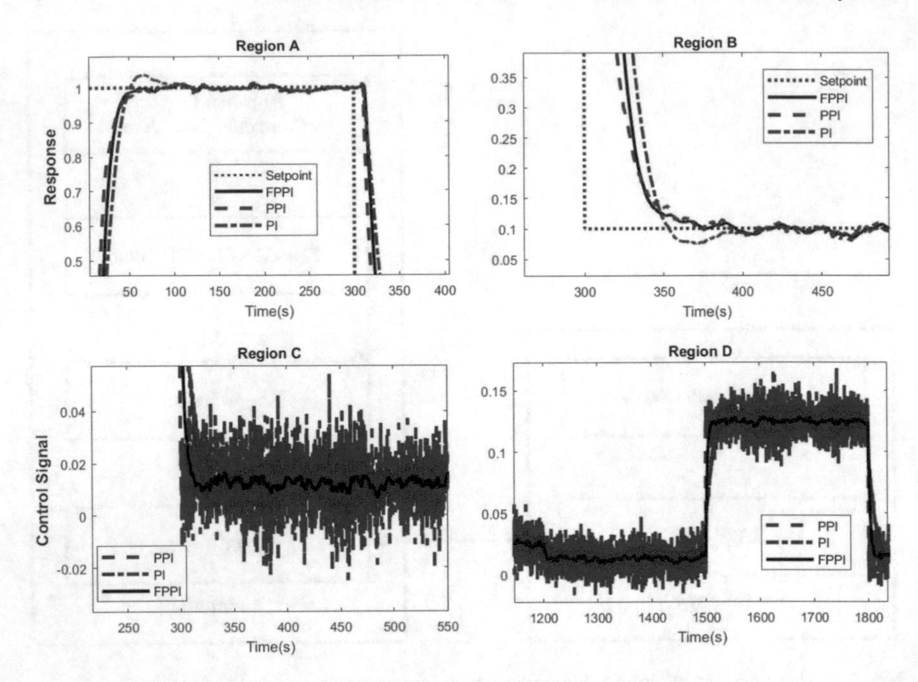

Fig. 2.29 Zoomed-in view of Fig. 2.28 for regions A, B, C and D

In order to validate the effectiveness of the design, a real-time simulation of the first order model Eq. (2.34) has been conducted with the WH-HILS using real-time delay obtained from the mote with mac address '00170D000030045B'. The experimental set-up is shown in Fig. 2.25. The results of the hardware in the loop simulation are shown in Fig. 2.28. The zoomed in for the regions of interest are shown in Fig. 2.29. The experiment has been conducted for a period of 2000s. By carefully observing Fig. 2.28, it can be seen that the proposed FPPI method is better than both PI and PPI controllers in terms of noise attenuation and good setpoint tracking. Similarly, by closely observing Fig. 2.29, the ability of the proposed approach to attenuate the effect of the noise on the control signal (regions C and D) as compared the other controllers is visible. The result here followed a similar pattern to the results of the pure simulation.

2.5 Summary

In this chapter, predictive PI controller has been adapted for use in wireless networked environment. The structure of the adapted controller has also been modified through the use of filter. This is in attempt to mitigate the effect of noise as demonstrated through simulation results. As part of an attempt to justify the controller, robustness and stability analysis has been provided. Parameters of the improved controller has also been tuned using our proposed hybrid algorithm (see Chap. 5). It has been observed from the results presented that, the approach can indeed be used in an environment characterised by stochastic network delay and noise.

References

1. Normey-Rico, J.E., Camacho, E.F.: Dead-time compensators: a survey. Control Eng. Pract. **16**(4), 407–428 (2008)
2. Hassan, S.M., Ibrahim, R., Saad, N., Asirvadam, V.S., Chung, T.D.: Setpoint weighted WirelessHART networked control of process plant. In: 2016 IEEE International Instrumentation and Measurement Technology Conference Proceedings, Taipei, Taiwan, 23–26 May 2016
3. Shinskey, F.G.: PID-deadtime control of distributed processes. Control Eng. Pract. **9**(11), 1177–1183 (2001)
4. Tan, K.K., Tang, K.Z., Su, Y., Lee, T.H., Hang, C.C.: Deadtime compensation via setpoint variation. J. Process Control **20**(7), 848–859 (2010)
5. Hägglund, T.: A predictive PI controller for processes with long dead times. IEEE Control Syst. Mag. **12**(1), 57–60 (1992)
6. Larsson, P., Hägglund, T.: Comparison between robust PID and predictive PI controllers with constrained control signal noise sensitivity. IFAC Proc. Vol. **45**(3), 175–180 (2012)
7. Hassan, S.M., Ibrahim, R., Saad, N., Asirvadam, V.S., Chung, T.D.: Predictive PI controller for wireless control system with variable network delay and disturbance. In: 2016 2nd IEEE International Symposium on Robotics and Manufacturing Automation, Ipoh, Perak, Malaysia, 25–27 September 2016
8. Larsson, P.O., Hägglund, T.: Control signal constraints and filter order selection for PI and PID controllers. In: Proceedings of the 2011 American Control Conference, Philadelphia, PA, USA, June 29–July 1 2011
9. Segovia, V.R., Hägglund, T., Åström, K.J.: Measurement noise filtering for PID controllers. J. Process Control **24**(4), 299–313 (2014)
10. Huba, M., Bélai, I.: Experimental evaluation of a DO-FPID controller with different filtering properties. IFAC Proc. Vol. **47**(3), 198–203 (2014)
11. Huba, M.: Filter choice for an effective measurement noise attenuation in PI and PID controllers. In: 2015 IEEE International Conference on Mechatronics, Beijing, China, 2–5 August 2015
12. Normey-Rico, J.E., Bordons, C., Camacho, E.F.: Improving the robustness of dead-time compensating PI controllers. Control Eng. Pract. **5**(6), 801–810 (1997)
13. Ribić, A.I., Mataušek, M.R.: A new predictive PI controller with additonal filtering. IFAC Proc. Vol. **45**(3), 489–494 (2012)
14. Huba, M.: Comparing 2DOF PI and predictive disturbance observer based filtered PI control. J. Process Control **23**(10), 1379–1400 (2013)
15. Arousi, F., Schmitz, U., Bars, R., Haber, R.: Robust predictive PI controller based on first-order dead time model. IFAC Proc. Vol. **41**(2), 5808–5813 (2008)
16. De Biasi, M., Snickars, C., Landernäs, K., Isaksson, A.J.: Simulation of process control with WirelessHART networks subject to clock drift. In: IEEE 2008 COMPSAC, pp. 1355–1360 (2008)
17. Chung, T.D., Ibrahim, R.B., Asirvadam, V.S., Saad, N.B., Hassan, S.M.: Adopting EWMA filter on a fast sampling wired link contention in WirelessHART control system. IEEE Trans. Instrum. Meas. **65**(4), 836–845 (2016)
18. Larsson, P.O., Hägglund, T.: Robustness margins separating process dynamics uncertainties. In: 2009 European Control Conference, Budapest, Hungary, 23–26 August 2009
19. Simon, D.: Analyzing control system robustness. IEEE Potentials **21**(1), 16–19 (2002)
20. Normey-Rico, J.E.: Control of Dead-Time Processes. Springer Science & Business Media, New York (2007)

21. Huang, G., Akopian, D., Chen, C.P.: Measurement and characterization of channel delays for broadband power line communications. IEEE Trans. Instrum. Meas. **63**(11), 2583–2590 (2014)
22. Ren, W., Steurer, M., Baldwin, T.L.: Improve the stability and the accuracy of power hardware-in-the-loop simulation by selecting appropriate interface algorithms. IEEE Trans. Ind. Appl. **44**(4), 1286–1294 (2008)
23. Ogan, R.T.: Hardware-in-the-loop simulation. Modeling and Simulation in the Systems Engineering Life Cycle. Springer, London (2015)
24. Sheng, S., Sun, C.: An adaptive attitude tracking control approach for an unmanned helicopter with parametric uncertainties and measurement noises. Int. J. Control Autom. Syst. **14**(1), 217–228 (2016)
25. Schlager, M.: Hardware-in-the-Loop Simulation. VDM Verlag, Riga (2008)

Chapter 3
WirelessHART Networked Set-Point Weighted Controllers

3.1 Introduction

This chapter presents the development of SW and adaptive SW controller for WHNCS. The SW technique is a powerful and simple method based on the feed-forward strategy. The advantage of this method is that it can be employed to improve systems performance with respect to setpoint tracking ability and disturbance rejection capability. The design gives a two-degree-of-freedom controller in which the first degree of freedom gives good setpoint response while the second degree of freedom gives good disturbance rejection as well as good robustness to model mismatch and other uncertainties. This type of controller adequately falls within the class of controllers to solve the problem of stochastic delay, random noise, and uncertainties.

The first part of this chapter will review some setpoint weighting based two degree of freedom (2-DoF) control strategies and attempt for their implementation for wireless networked systems. Then, the design of the setpoint weighting controller for WirelessHART networked system will be presented starting with a simple setpoint weighting structure. Robustness and stability analysis of the proposed controller will also be presented followed by simulation study. The later part of the chapter will use fuzzy tuner to adapt the developed controller to wider variation of the network delay. In that regards, simulation study will also be presented.

3.2 Brief Review of Setpoint Weighting Strategies

Implementation of 2-DoF control through SW technique for delayed systems has been attempted by many researchers [1–7]. Foremost, the strategies focus on avoiding the complexity of model-based controllers while maintaining the simplicity of the PID controller. The other feature of these strategies is that they address the problem of the gain range associated with PID when used in the delayed environment. Lastly,

S. M. Hassan et al., *Hybrid PID Based Predictive Control Strategies for WirelessHART Networked Control Systems*, Studies in Systems, Decision and Control 293,
https://doi.org/10.1007/978-3-030-47737-0_3

the implementation of the SW structure can be achieved without altering the closed loop stability because the structure lies outside the closed loop.

On the 2-DoF control strategies, for instance, the SW strategy reported in [2] involves the use of a flexible structure which treats separately the setpoint and the process output. This is achieved by incorporating two tunable gains in the proportional and derivative terms of the PID controller. The aim of this method is for the controller to be robust to setpoint changes and load disturbance. The problem with this method is that it does not solve the problem of gain limitation. This is because the proportional gain β introduced is still limited usually within the range of 0 and 1 (i.e. $0 \leq \beta \leq 1$). Furthermore, the gain associated with the derivative term is set to 0 to avoid transient due to setpoint changes. Another disadvantage of this method is that part of the structure lies within the closed loop of the original PID. Thus using it may require some modification to the existing structure.

The use of fuzzy logic has been proposed to tune setpoint weight of PID controller in both [8, 9]. Here, while the weighting term associated with the proportional action of the controller (β) is tuned via fuzzy logic, the load regulation performance is achieved by preserving or improving on the Ziegler âĂŞ Nichols formula. The drawback of this method is that it is based on the previous method in [2] that has a limited gain range between 0 and 1. Thus, this method does not extend significantly the range of gain used for the controller. It should be noted that with this method, any gain above unity will cause large overshoots and oscillation. Another drawback of this method is that the use of fuzzy logic imposes additional task of having to tune the fuzzy logic parameters. In [4] a setpoint weighted multivariable PID controller was tuned using bilinear matrix inequalities (BMI) optimization. Again, this method is based on the method proposed in [2] thus, has the same limitation of gain range and structure. Attempt to increase the gain (i.e. $\beta > 1$) of weight associated with the proportional term of the PID controller was made in [3]. Here, the selection criteria for the proportional weight β depends on the design parameter T_c and the PI controller gain K_c. Although this method has allowed the use of $\beta > 1$, it has not significantly increased the gain range as the maximum value of β achieved is limited to less than 3. In that case, the high gain increases the sluggishness of the response. Online dynamic SW schemes using linear relation, fuzzy and sigmoid function to dynamically tune β were proposed in [5, 10, 11] respectively. These methods still retain the limitation of restricting β to less than 1.

In [6, 7], the use of a filter with the structure of first-order plus dead time (FOPDT) in a set point weighing arrangement to achieve 2âĂŞDoF control is reported. Notably, the complete structure is similar to the that reported in [1] with the exception that the entire SW block configuration of the two are different. In the structure, the SW block is outside the closed loop control setup. However, no attempt has been made to implement the 2âĂŞDoF control for the WirelessHART networked environment until when the structure reported in [1] was adopted for WirelessHART control and reported in our initial work in [12]. The structure provides for flexibility in the gain range which the earlier methods discussed above did not.

3.3 Design of Setpoint Weighting Controller for WirelessHART Networked Control System

This section will begin by presenting the most simple setpoint weighting function. Then a general setpoint weighting function will be presented before adopting it for wireless Hart networked environment. This will be done in order to ease the understanding of the reader. Under this subsection, comprehensive procedures for the design and selection of parameters of both the setpoint weighting function as well as the associated PI controller will be discussed. The robustness analysis of the controller will be used to recap the section.

3.3.1 Simple SW Function

Consider the SW control structure of Fig. 3.1. Assuming $f_r(s)$ is the simple SW function given as

$$f_r(s) = \frac{\tilde{r}(s)}{r(s)} = 1 + \tilde{G}_{yr}(s)(e^{\tilde{\tau}s} - 1) \tag{3.1}$$

where the estimate of the delay is $\tilde{\tau}$ and the desired closed loop response free from delay is $\tilde{G}_{yr}(s)$.

Thus, the close-loop transfer function is

$$\frac{y(s)}{r(s)} = \frac{G_c(s)P(s)e^{-\tau s}}{1 + G_c(s)P(s)e^{-\tau s}} \left[1 + \tilde{G}_{yr}(s)(e^{\tilde{\tau}s} - 1)\right] \tag{3.2}$$

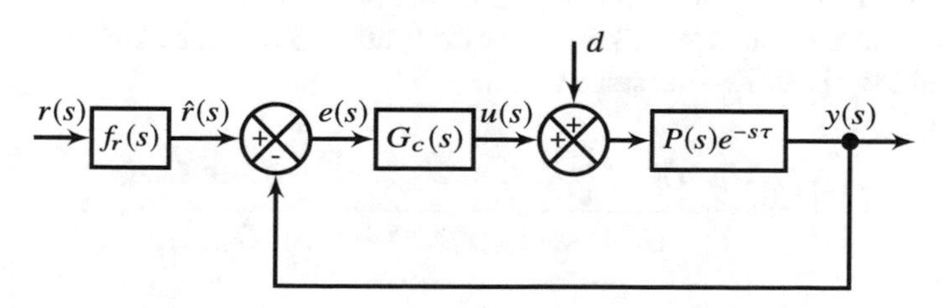

Fig. 3.1 SW structure without the WirelessHART network

If $G_{yr}(s) = \frac{G_c(s)P(s)}{1+G_c(s)P(s)}$, then Eq. (3.2) can be written as

$$\frac{y(s)}{r(s)} = \frac{G_{yr}(s)e^{-\tau s}\left[1 + \tilde{G}_{yr}(s)e^{\tilde{\tau}s} - \tilde{G}_{yr}(s)\right]}{1 + G_{yr}(s)e^{-\tau s} - G_{yr}(s)} \quad (3.3)$$

$1 + G_{yr}(s)e^{-\tau s} - G_{yr}(s) = 0$, the denominator of Eq. (3.3) can be shown to have the same solution as $1 + G_c(s)P(s)e^{-\tau s} = 0$ which is the characteristic equation of Fig. 3.1 when $f_r(s) = 0$.

It is assumed that $\tilde{\tau} = \tau$ and $G_{yr}(s) = \tilde{G}_{yr}(s)$. Thus, if $G_c(s)$ is designed such that the characteristics equation of Eq. (3.3) has real negative parts only, pole-zero cancellation on Eq. (3.3) can be done. Thus, the transfer function from $r(s)$ to $y(s)$ in Fig. 3.1 is

$$\frac{y(s)}{r(s)} = G_{yr}(s)e^{-\tau s} \quad (3.4)$$

where $G_{yr}(s) = \frac{G_c(s)P(s)}{1+G_c(s)P(s)}$ is the closed loop transfer function without the delay.

The relationship in (3.4) shows that the delay term $e^{-\tau s}$ is decoupled from the delay-free function $G_{yr}(s)$. This enables the design of $G_c(s)$ to be done with respect to the delay-free portion of the process $P(s)$. However, this approach still does not significantly extend the gain range of $G_c(s)$ for closed -loop stability. Therefore, a more general structure to is needed to extend the gain range.

3.3.2 General SW Function

To allow the use of a wider range of controller gain while still maintaining closed-loop stability, a generic SW function Eq. (3.5) was proposed in [1]. This permits the 2-DoF ability of both good setpoint tracking and disturbance rejection of the controller.

$$f_r(s) = \frac{\tilde{r}(s)}{r(s)} = G_r(s) + \tilde{G}_{yr}(s)\left(e^{\tilde{\tau}s} - G_r(s)\right) \quad (3.5)$$

where $G_r(s)$ is the setpoint regulating feed-forward controller.

In a similar fashion to Sect. 3.3.1, after integrating $G_r(s)$, the transfer function of the closed loop is written the same way as

$$\frac{y(s)}{r(s)} = \frac{\hat{G}_{yr}(s)e^{-\tau s}\left[G_r(s) + \tilde{G}_{yr}(s)e^{\tilde{\tau}s} - G_r(s)\tilde{G}_{yr}(s)\right]}{G_r(s) + \hat{G}_{yr}(s)e^{-\tau s} - G_r(s)\hat{G}_{yr}(s)} \quad (3.6)$$

If $\hat{G}_{yr}(s) = \frac{G_r(s)G_c(s)P(s)}{1+G_r(s)G_c(s)P(s)}$ is the desired closed loop transfer function with the higher gain, $\tilde{\tau} = \tau$ and $\hat{G}_{yr}(s) = \tilde{G}_{yr}(s)$, and after pole-zero cancellation on (3.6). Hence, the delay-free portion of the process $\hat{G}_{yr}(s)$ can be decoupled from the delay term $e^{-\tau s}$ as in (3.4). Such that the overall transfer function of the loop is written as

$$\frac{y(s)}{r(s)} = \hat{G}_{yr}(s)e^{-\tau s}. \tag{3.7}$$

3.3.3 SW Function for WHNCS

Consider the closed-loop WirelessHART network of Fig. 3.2 with a setpoint weighting function $f_r(s)$, the structure permits variation of the reference signal from $r(s)$ to $\tilde{r}(s)$. This allows for the 2-DoF ability of both good setpoint tracking and disturbance rejection of the controller. Note that the plant model $G(s) = P(s)e^{-s\tau}$ where plant deadtime is given as τ.

The SW function $f_r(s)$ is a general SW function [1] given as

$$f_r(s) = \frac{\tilde{r}(s)}{r(s)} = G_r(s) + \tilde{G}_{yr}(s)\big(e^{\tilde{\tau}s} - G_r(s)\big) \tag{3.8}$$

where $G_r(s)$ is the setpoint regulating feed-forward controller. The block diagram implementation of the function is shown in Fig. 3.3.

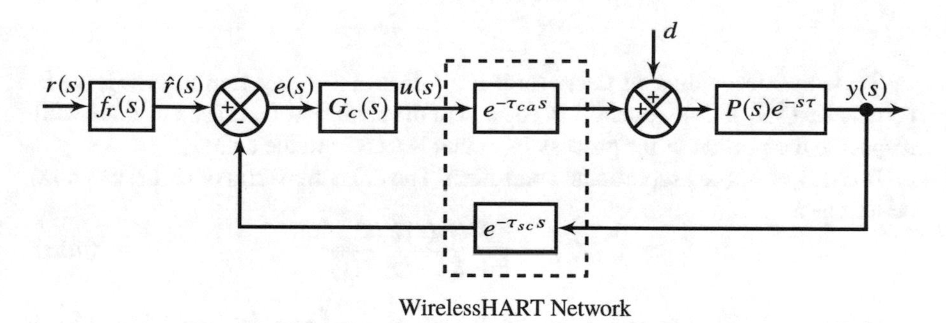

Fig. 3.2 SW structure with WirelessHART network

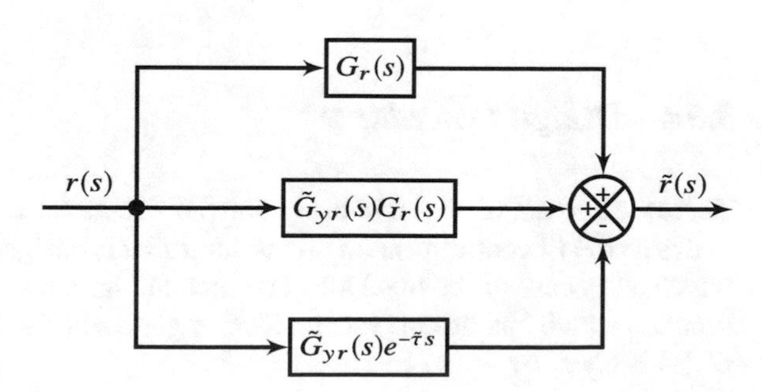

Fig. 3.3 Structure of the general SW function $f_r(s)$

The SW controller design for WHNCS is done by analyzing the diagram of Fig. 3.2 for closed loop control. The figure is based on the structure discussed in [1]. The transfer function from $y(s)$ to $r(s)$ assuming commutativity of the terms is

$$\frac{y(s)}{r(s)} = \frac{G_c(s)P(s)e^{-(\tau_{CA}+\tau)s}}{1+G_c(s)P(s)e^{-(\tau_{CA}+\tau_{SC}+\tau)s}} f_r(s) \tag{3.9}$$

Defining $\tau_1 = \tau_{CA} + \tau$ and $\tau_2 = \tau_{CA} + \tau_{SC} + \tau$, (3.9) can be written as

$$\frac{y(s)}{r(s)} = \frac{G_c(s)P(s)e^{-\tau_1 s}}{1+G_c(s)P(s)e^{-\tau_2 s}} f_r(s) \tag{3.10}$$

Substituting Eqs. (3.8) in (3.10) and defining $\hat{G}_{yr}(s) = \frac{G_r(s)G_c(s)P(s)}{1+G_r(s)G_c(s)P(s)}$ as the desired closed loop transfer function, the closed-loop transfer function of Fig. 3.2 can be written as

$$\frac{y(s)}{r(s)} = \frac{\hat{G}_{yr}(s)e^{-\tau_1 s}\left[G_r(s) + \tilde{G}_{yr}(s)e^{\tilde{\tau}s} - G_r(s)\tilde{G}_{yr}(s)\right]}{G_r(s) + \hat{G}_{yr}(s)e^{-\tau_2 s} - G_r(s)\hat{G}_{yr}(s)} \tag{3.11}$$

Assuming $\tilde{\tau} = \tau_2$ and $\hat{G}_{yr}(s) = \tilde{G}_{yr}(s)$ and pole-zero cancellation in (3.11), then

$$\frac{y(s)}{r(s)} = \hat{G}_{yr}(s)e^{-\tau_1 s} \tag{3.12}$$

Thus, the decoupling of delay term $e^{-\tau_1 s}$ from the delay free term $\hat{G}_{yr}(s)$ is revealed in (3.12). Consequently, the design of the controller $G_c(s)$ can be done with respect to the portion of the process $P(s)$ that is free from the delay.

If $G_r(s) = K$ (i.e proportional controller). The delay free term of (3.12) can now be written as

$$\hat{G}_{yr}(s) = \frac{KG_c(s)P(s)}{1+KG_c(s)P(s)} \tag{3.13}$$

Equation (3.13) is an indication that 2-DoF of good setpoint tracking and load regulation can be achieved while still permitting the use of high stable gain K.

3.3.4 Design Procedures

There are two major phases of designing SW function for WHNCS. The first phase is to design the PI controller $G_c(s)$. This controller is designed for good load regulation which takes care of the first DoF. The second phase takes care of the design of SW function which handles the second DoF of good setpoint tracking. The PI controller $G_c(s)$ is given by

$$G_c(s) = K_c\left(1 + \frac{1}{T_i s}\right) \tag{3.14}$$

where,

$$K_c = \frac{T}{2K_p(L + \Delta L)} \tag{3.15}$$

$$T_i = T \tag{3.16}$$

The gain K_c and time constant T_i of the PI controller given in Eqs. (3.15) and (3.16) are designed based on modification of the method proposed by [14]. The term ΔL should be chosen between 5–20% of the total delay L.

The design ensures robustness to load disturbances while delivering around 5% overshoot. The design of $f_r(s)$ involves deigning two components $G_r(s)$ and $\hat{G}_{yr}(s)$. The first component $G_r(s)$ can be chosen simply as K if not much information on the process is available. However, if there is enough information of the process to be controlled, it can be designed as Eq. (3.17).

$$G_r(s) = \frac{K G_c(s)^{-1} P(s)^{-1}}{B_c(s)} \tag{3.17}$$

where $B_c(s)$ is the denominator of $G_c(s)$ when expressed as $\frac{A_c(s)}{B_c(s)}$.

The desired closed loop transfer function is given as

$$\hat{G}_{yr}(s) = \frac{1}{B_c(s)/K + 1} \tag{3.18}$$

$B_c(s) = T_i s$ if $G_c(s)$ is a PI or PID controller. Hence, (3.18) can be written as

$$\hat{G}_{yr}(s) = \frac{1}{T_i s/K + 1} \tag{3.19}$$

where K is chosen such that the desired closed loop response is met.

It should be noted that the design for $G_r(s)$ in the case of second and higher order systems leads to a transfer function with more zeros than poles, this can be corrected by appending fast poles to the transfer function of $G_r(s)$. Usually, the design for second order system leads to one more zero and that for third order system leads to two more zeros than poles in the transfer function of $G_r(s)$. Thus appending poles should be added to counter the corresponding extra zeros. Alternatively, $G_r(s)$ can be designed simply as a gain K as explained above.

3.3.5 Robustness Analysis

Under acceptable conditions, the SW for the wireless control system with the assumptions that $\tilde{\tau} = \tau_2$ and $\hat{G}_{yr}(s) = \tilde{G}_{yr}(s)$ will result in improved setpoint tracking the performance of the system. It should be noted that since the SW term $f_r(s)$ is outside the closed loop as shown in Fig. 3.2, it does not affect the stability and robustness of the system in as much as it does not add an unstable pole to the system. This conforms to the robustness analysis presented in [1] for 2-DoF controllers. However, if the acceptable conditions are not met i.e. $\tilde{\tau} \neq \tau_2$ and $\hat{G}_{yr}(s) \neq \tilde{G}_{yr}(s)$, the robustness and setpoint tracking performance will be affected.

A general condition to maintain robust tracking performance for systems with long deadtime is presented in [1]. Thus, accordingly, we adopt and modify some of these conditions to suit the WirelessHART networked control environment. The tracking error of Fig. 3.2 is given by

$$e_{rr}(s) = \tilde{r}(s) - y(s)e^{-\tau_{SC}s} \tag{3.20}$$

It follows from Eq. (3.12) that the output can be expressed as $y(s) = r(s)\hat{G}_{yr}(s)e^{-\tau_1 s}$ also from (3.8), $f_r(s) = \frac{\tilde{r}(s)}{r(s)}$. Thus (3.20) can be written as

$$e_{rr}(s) = r(s)\Big(f_r(s) - \frac{f_r(s)}{\hat{f}_r(s)}\hat{G}_{yr}(s)e^{-\tau_2 s}\Big) \tag{3.21}$$

where the setpoint weighing function under nominal condition $\tilde{\tau} = \tau_2$ and $\hat{G}_{yr}(s) = \tilde{G}_{yr}(s)$ is $\hat{f}_r(s) = G_r(s) + \hat{G}_{yr}(s)\big(e^{\tilde{\tau}s} - G_r(s)\big)$.

Thus, if these nominal conditions hold, the setpoint term is expressed as $f_r(s) = \hat{f}_r(s)$ and the error in (3.21) becomes

$$e_{rr}(s) = r(s)\Big(f_r(s) - \hat{G}_{yr}(s)e^{-\tau_2 s}\Big) \tag{3.22}$$

When there is a deviation from nominal conditions, the deviation can be modeled as either an additive or multiplicative uncertainty.

The additive uncertainty in the frequency domain (i.e. when $s = i\omega$) is represented as $f_r(i\omega) = \hat{f}_r(i\omega) + l_a(i\omega)$. Hence, the sensitivity function derived from Eq. (3.21) should satisfy the following condition:

$$\Big|\big(\hat{f}_r(i\omega) + l_a(i\omega)\big)\Big(1 - \frac{1}{\hat{f}_r(i\omega)}\hat{G}_{yr}(i\omega)e^{-\tau_2 i\omega}\Big)\alpha\Big| < 1 \tag{3.23}$$

where α is the robust performance weight and l_a is the additive uncertainty term. Consequently, (3.24) can be written in the form

$$\left(\left|\hat{f}_r(i\omega)\right| + \left|l_a(i\omega)\right|\right)\left|\left(1 - \frac{1}{\hat{f}_r(i\omega)}\hat{G}_{yr}(i\omega)e^{-\tau_2 i\omega}\right)\alpha\right| < 1 \tag{3.24}$$

with $|l_a(i\omega)| \leq \bar{l}_a(\omega)$ being the bound on the uncertainty. Therefore, the SW function should be designed to satisfy Eq. (3.24) for robust performance.

Similarly, when multiplicative uncertainty is considered, the deviation can be represented as $f_r(i\omega) = \hat{f}_r\big(i\omega)(1 + l_m(i\omega)\big)$. The uncertainty function also derived from (3.21) should satisfy the following for robust performance:

$$\left|\hat{f}_r(i\omega)\right|\left(1 + \left|l_m(i\omega)\right|\right)\left|\left(1 - \frac{1}{\hat{f}_r(i\omega)}\hat{G}_{yr}(i\omega)e^{-\tau_2 i\omega}\right)\alpha\right| < 1 \tag{3.25}$$

where l_m is the multiplicative uncertainty term. Hence, the multiplicative uncertainty bound condition is $|l_m(i\omega)| \leq \bar{l}_m(\omega)$. Likewise the SW function should be designed to satisfy (3.25) for robust performance if multiplicative uncertainty is considered.

3.4 Performance of Proposed SW Controller

This section presents results and analyses the performance of both SW and Fuzzy adaptive SW strategies. For the SW strategy, the comparison will be made with PI and Smith predictor, while the FASW will be compared against SW and PI. This is because both the proposed SW and the Smith predictor are PI-based control strategies [1, 13].

The SW and PI controller parameters for system models given in Eqs. (2.34), (2.35), (2.36) and (2.37) are shown in Table 3.1. These parameters are used where necessary in the proposed SW strategy, the PI and in the Smith predictor controller. In the table, K_{C1} is the PI gain used in the SW strategy, whereas K_{C2} is PI controller gain according to the design in [14]. The PI design ensures not more than 5% overshoot with good load regulation. In order to achieve faster recovery from the effect of disturbance offset without compromise to overshoot with the SW controller, the value of K_{C1} should be chosen between 85–95% of the calculated value as recommended in [1]. To evaluate the performance of the proposed SW method, plants considered are simulated to a unit step signal. A disturbance signal of magnitude 0.5 with a step time of 200 s is injected at the input of all plants. All simulations are done using measured network delays of Fig. 3.4 and measurement noise. The performance evaluation is based on the comparison of the integral absolute error (IAE), rise time (response speed), settling time and percentage overshoot. Furthermore, the proposed SW controller is tested for its ability to track changing setpoint for all systems, while its robustness to parametric model mismatch is tested for the first-order system.

Table 3.1 SW and PI parameters

Plant	Parameter				
	$G_r(s)$	$\hat{G}_{yr}(s)$	K_{C1}	K_{C2}	T_i
$G_1(s)$	25.83	$\frac{8}{9.13s+8}$	0.0387	0.0445	2
$G_2(s)$	$\frac{10.8(s+1)}{(1.3s+1)}$	$\frac{1}{1.3s+1}$	0.0789	0.0929	1.3
$G_3(s)$	$\frac{10(s+1)}{(2s+1)}$	$\frac{1}{2s+1}$	0.0954	0.1061	2
$G_4(s)$	$\frac{12.274(0.016s^4+0.234s^3+1.094s^2+1.87s^4+1)}{(1.5^4+5.5s^3+7.5s^2+4.5s^4+1)}$	$\frac{1}{1.5s+1}$	0.0895	0.0942	1.5

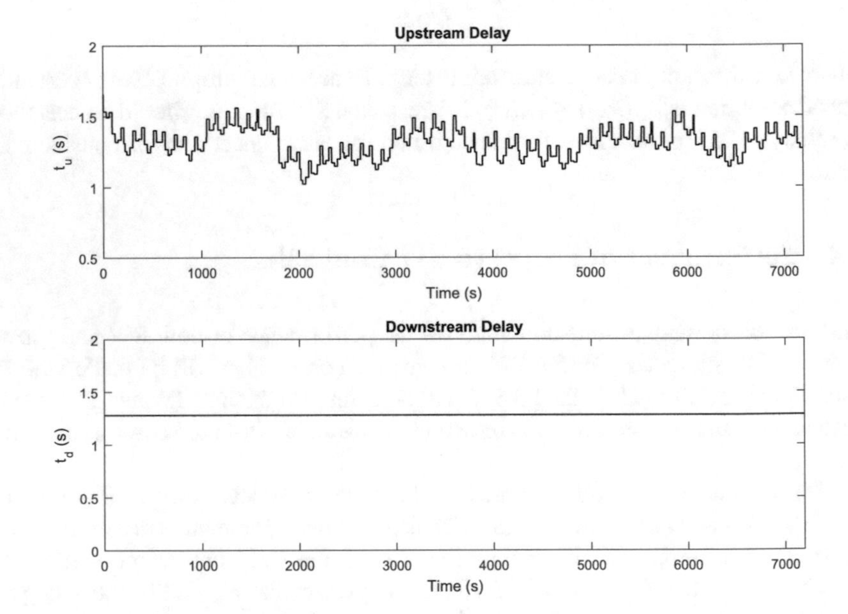

Fig. 3.4 Measured network delay profile

3.4.1 *First Order System*

By carefully observing Fig. 3.5 and Table 3.2, it can be easily seen that the time domain performance of the proposed SW controller for setpoint tracking and disturbance rejection is better compared to the performance of both PI and Smith predictor. The rise time of the plant with the proposed SW at around 5 s is almost three and six times faster than those of the PI and Smith predictor respectively. The settling time with the proposed SW follows a similar pattern to the response time. In terms of overshoot, the proposed SW fares favorably compared to the Smith predictor while much better than the PI with a value of 5.7%. Smaller value IAE is recorded for the proposed method compared to those of PI and Smith predictor methods. Furthermore, the proposed SW recovered faster and without overshoot from the effect of

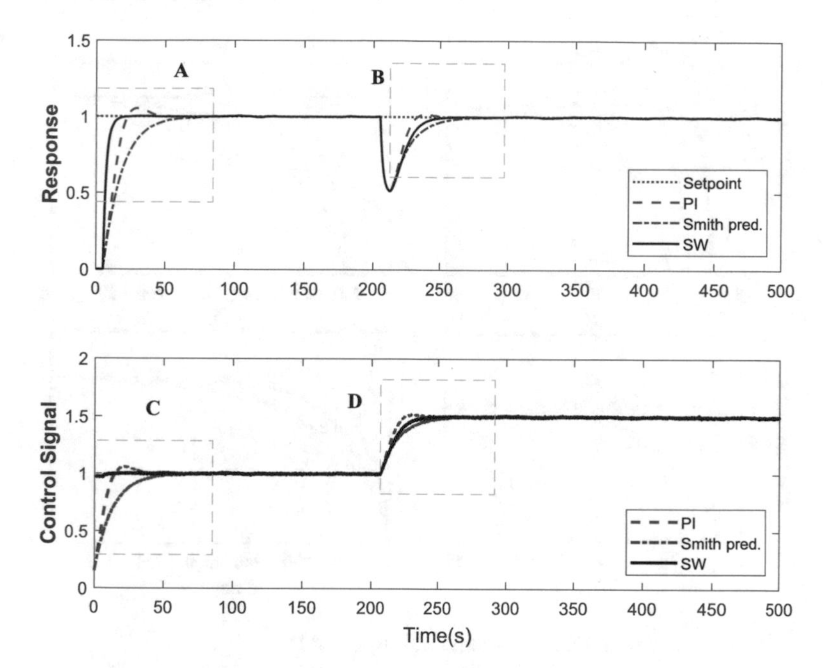

Fig. 3.5 Response of first order system to various controllers with disturbance

Table 3.2 Control performance of first order system

Controller	System's performance			
	Rise time (s)	Settling time (s)	Overshoot (%)	IAE
SW	4.9021	14.8910	0.6770	186.7814
PI	12.1877	40.6986	5.7198	229.4248
Smith Predictor	28.0783	55.1527	0.5505	307.2321

disturbance when compared to PI controller. Compared to the Smith predictor, the recovery is still faster as observed from the figure.

The fast controller action of the proposed SW to the step change can also be noticed from Fig. 3.5 as against the other two controllers. The zoomed-in view of regions of interest A, B, C and D in Fig. 3.5 is presented in Fig. 3.6.

Figure 3.7 shows the robustness of the proposed method to changing setpoint. Here, the ability of the proposed approach to track changing setpoint with minimal error is demonstrated. Also, the SW controller is able to track the setpoint faster without overshoot or undershoot compared to the PI controller. Although the Smith predictor produces no overshoot, its response is sluggish compared to the other two controllers.

In Fig. 3.8, the robustness of the SW controller to parametric model mismatch is presented. For this purpose, parameters of the model (i.e. time constant and the gain) are varied 5% above and 5% below their nominal value while the controller

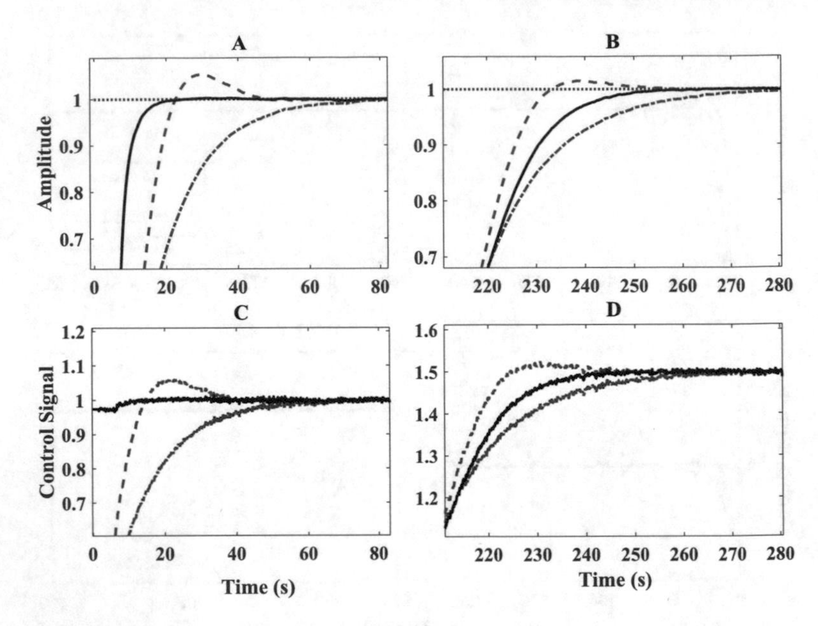

Fig. 3.6 Zoomed-in view of Fig. 3.5 for regions A, B, C and D

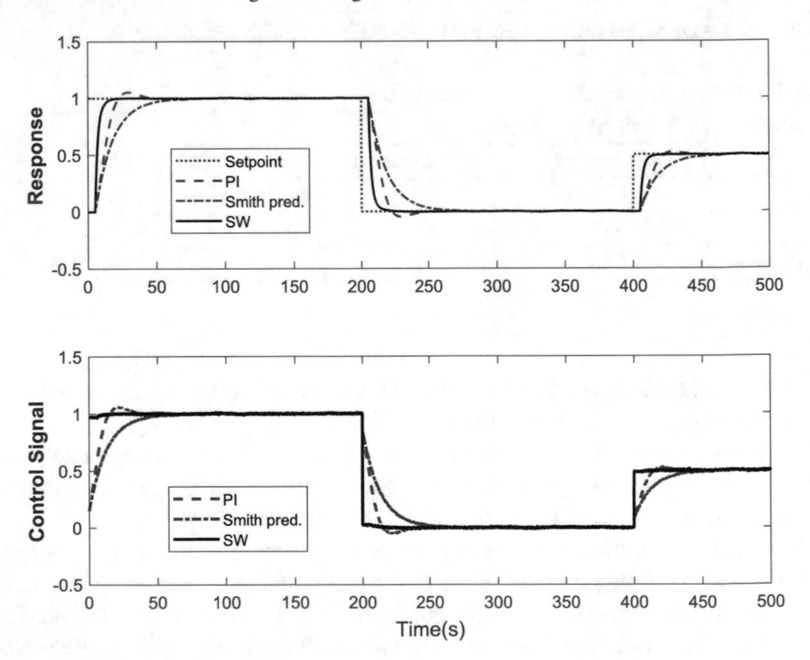

Fig. 3.7 Changing setpoint tracking of the first order system with disturbance

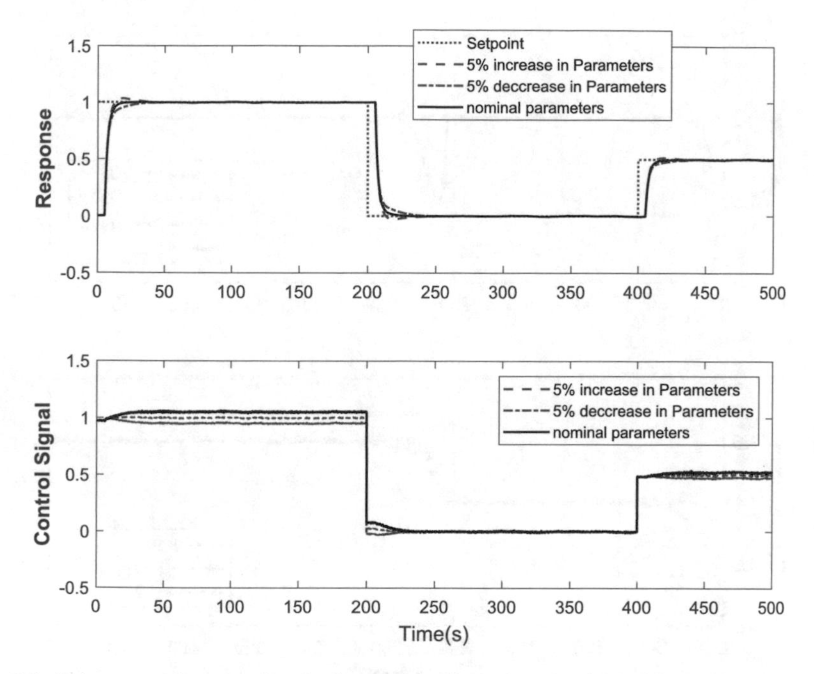

Fig. 3.8 Robustness of proposed method to model mismatch and variable delay, first order system

parameters remain unchanged. Simulation results with these variations are compared to that of the original values. It can be clearly seen from these graphs that despite the variations, no significant change is observed in terms of the ability of the controller to track the change in setpoint. This clearly reaffirmed the assertion that the controller can be employed in an environment characterized by uncertainties. It also confirms that exact model is necessary but not compulsory for the implementation.

3.4.2 Second Order System

The simulation results for this plant are shown in Fig. 3.9 and the zoomed-in of the various regions of interest is given in Fig. 3.10. The information regarding these figures is reported numerically in Table 3.3. In consonant with the results obtained for the first order system, similar trends are observed for the second order system. The response with proposed SW at 10 s is faster than those of PI and Smith predictor at 12.7 and 31 s respectively. While the plant with the SW controller settled at around 24 s, the respective settling times of PI and Smith predictor are around 44 and 61 s. This indicates that the proposed approach settled much faster than the other controllers. It is observed that the proposed SW approach produced overshoot of around 1% as against 0.55 and 7.2% of Smith predictor and PI respectively. The Smith predictor, despite having lower overshoot, is the slowest in both response and settling times compared

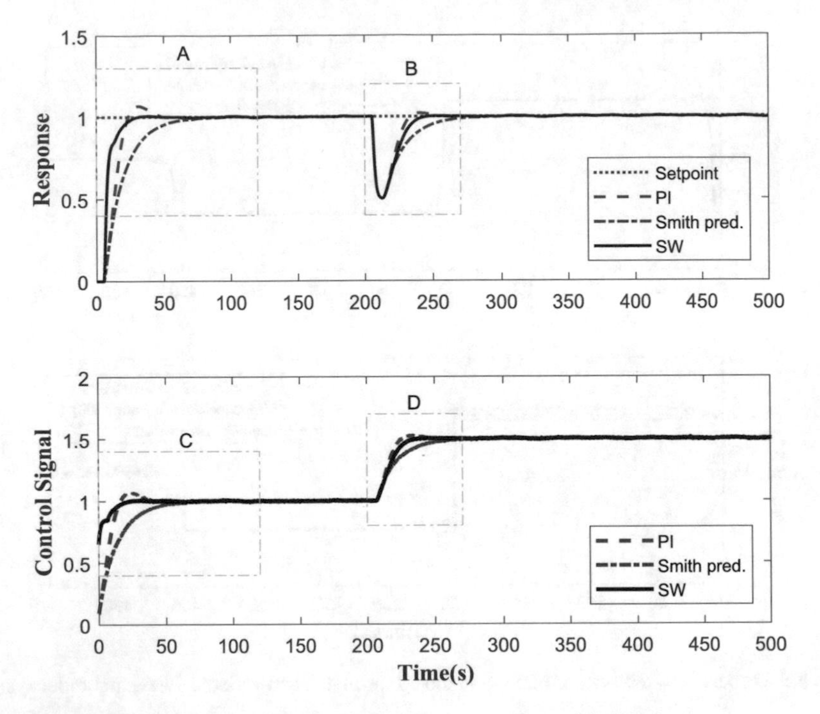

Fig. 3.9 Response of second order system to various controllers with disturbance

to both proposed SW and PI. Furthermore, the proposed SW controller produced the least IAE of around 204 as against 249.4 and 334 for PI and Smith predictor respectively. Observing Fig. 3.10, it is seen that, the recovery from disturbance effect of the proposed SW is also better than both PI and Smith predictor. This is because the proposed SW recovered without overshoot even though a bit slower than that of PI. The recovery of the approach is also much faster than the Smith predictor despite both recovered without overshoot.

Observing the action of various controllers, the proposed SW approach gives a more aggressive signal starting at 0.7 as against the PI and Smith predictor both starting at around 0 (see region C). However, in region D, the SW produced moderate action that is in-between those of faster PI and slower Smith predictor. This is to avoid overshoot while recovering from disturbance effect.

To analyse the performance of SW controller for changing setpoint tracking on the second order system, we observe the results of Fig. 3.11. From the results, it can be clearly seen that the SW controller outperformed the other two controllers compared (i.e. PI and Smith predictor). This can further be corroborated when various controller actions are observed in the same figure. It can be seen that the proposed controller is more aggressive at the point of setpoint change than both PI and Smith predictor. The results here also followed a similar trend to the first order system.

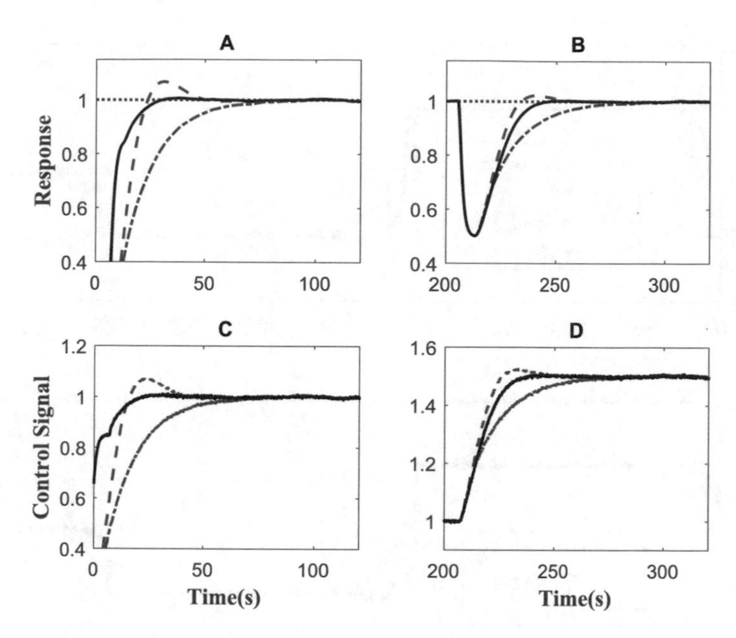

Fig. 3.10 Zoomed-in view of Fig. 3.9 for regions A, B, C and D

Table 3.3 Control performance of second order system

Controller	System's performance			
	Rise time (s)	Settling time (s)	Overshoot (%)	IAE
SW	10.0123	24.1357	1.0374	204.3250
PI	12.7245	44.6820	7.2109	249.3840
Smith Predictor	31.0791	61.6465	0.5563	334.0003

3.4.3 *Third Order System*

Simulation results for the third order system are shown in Fig. 3.12. The zoomed-in of regions A, B, C and D in the figure is given in Fig. 3.13. In Table 3.4, numerical results of the two figures are given. The trend of the result here is in conformity to those of first and second order systems. The response with proposed SW is almost three (3) times faster than that of PI controller and more than five (5) times that of Smith predictor at 6.89, 18.31 and 37.89 s respectively. The SW controller settled at around 22.5 s. This is less than half the settling time of PI and around one-third of the Smith predictor. This shows that the proposed approach settled much faster than the other controllers. While both the proposed SW controller and the Smith predictor produce respective overshoots of 0.77 and 0.60%, the PI controller produces overshoot of 2.62%. The proposed SW approach also produced the least IAE of around 240.50 as against 302.97 and 404.29 of PI and Smith predictor respectively. The recovery

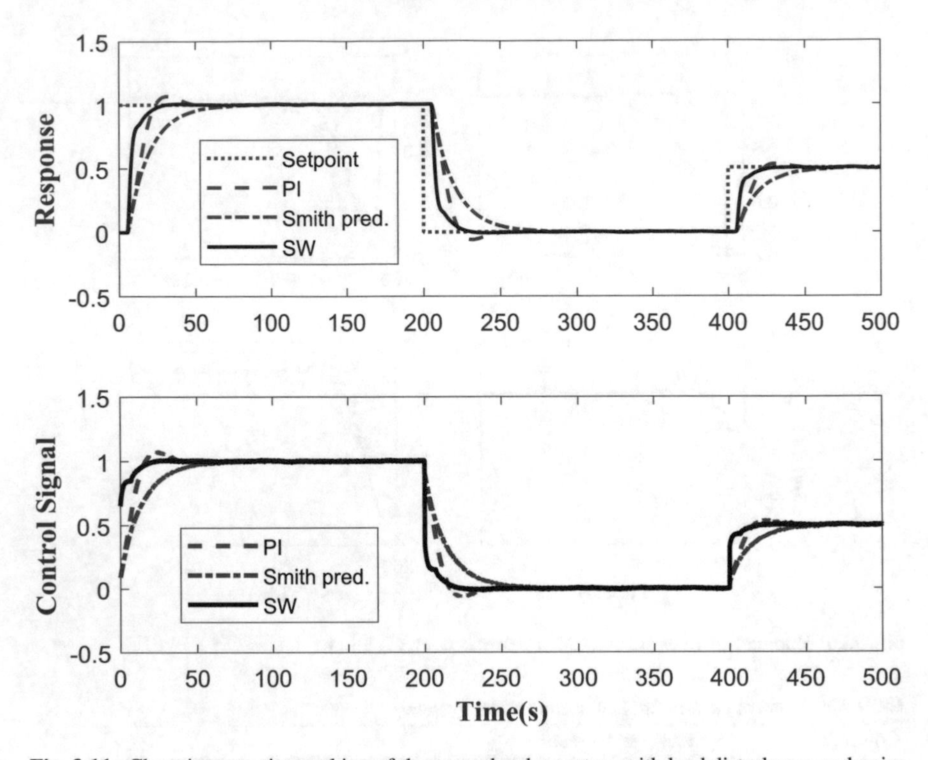

Fig. 3.11 Changing setpoint tracking of the second order system with load disturbance and noise

from disturbance effect of the proposed method is smooth without overshoot when compared to the little overshoot produced by the PI. Furthermore, the recovery of the proposed approach is faster than that of the Smith predictor even though the latter did not produce overshoot.

Observing the various control signals, the proposed approach produces more aggressive signal than other controllers during step change (see regions A and C of Fig. 3.13). An intermediate control signal between that of PI and Smith predictor is produced by the SW controller at the point of disturbance. This is to avoid overshoot while achieving fast disturbance recovery (see regions B and D of Fig. 3.13).

A careful observation of the results of Fig. 3.14 reveals the performance of SW controller for tracking changing setpoint on third order system. From the results, it can be clearly seen that the proposed approach outperformed the other two controllers compared i.e. PI and Smith predictor. This can further be corroborated when various controller actions are observed in the same figure, where it is seen that the proposed controller is more aggressive the others.

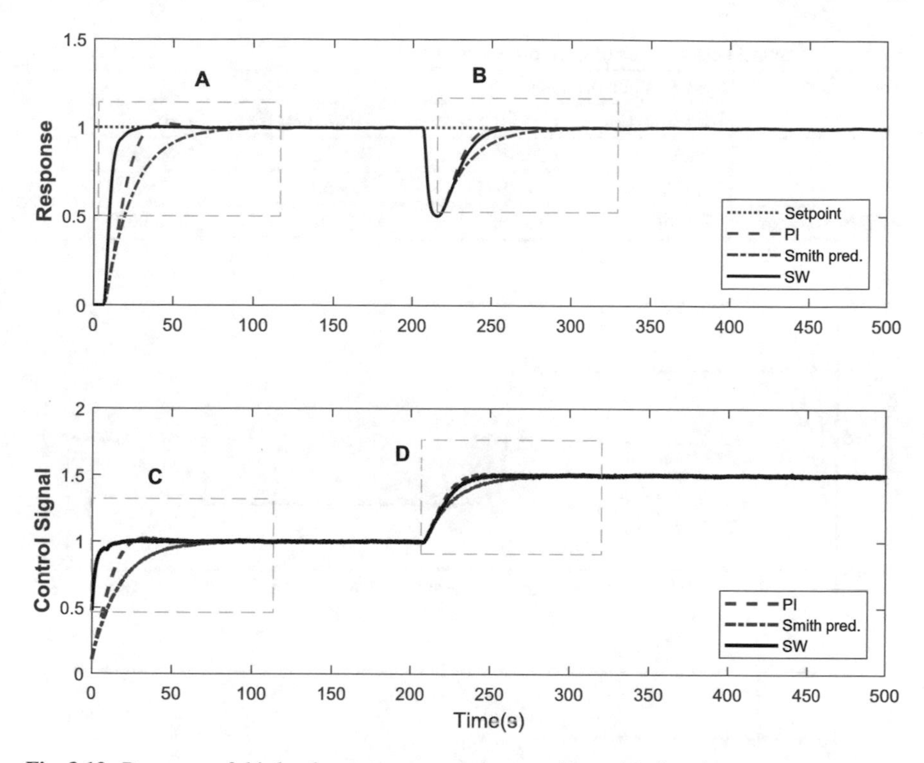

Fig. 3.12 Response of third order system to various controllers with disturbance

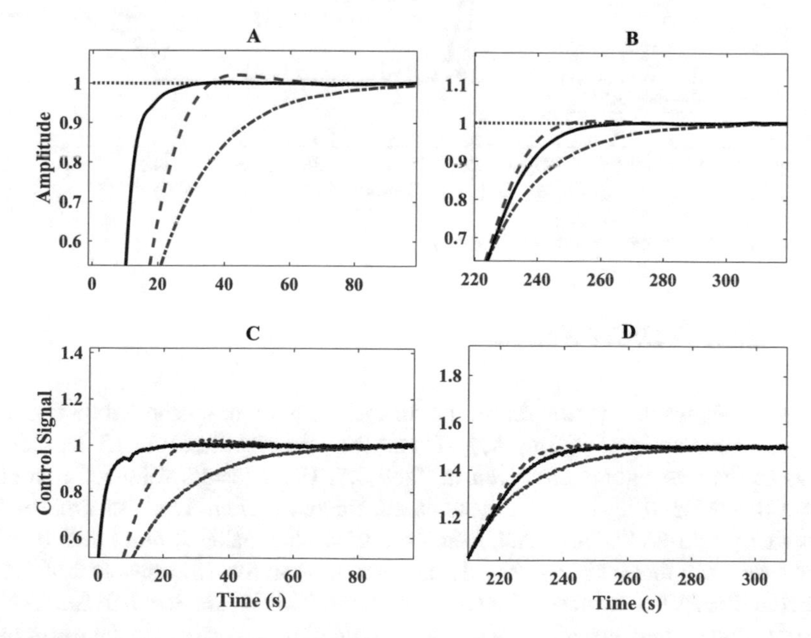

Fig. 3.13 Zoomed-in view of Fig. 3.12 for regions A, B, C and D

Table 3.4 Control performance of third order system

Controller	System's performance			
	Rise time (s)	Settling time (s)	Overshoot (%)	IAE
SW	6.8902	22.4816	0.7712	239.4999
PI	18.3065	51.3941	2.6163	302.9715
Smith Predictor	37.8866	74.7931	0.5992	404.2869

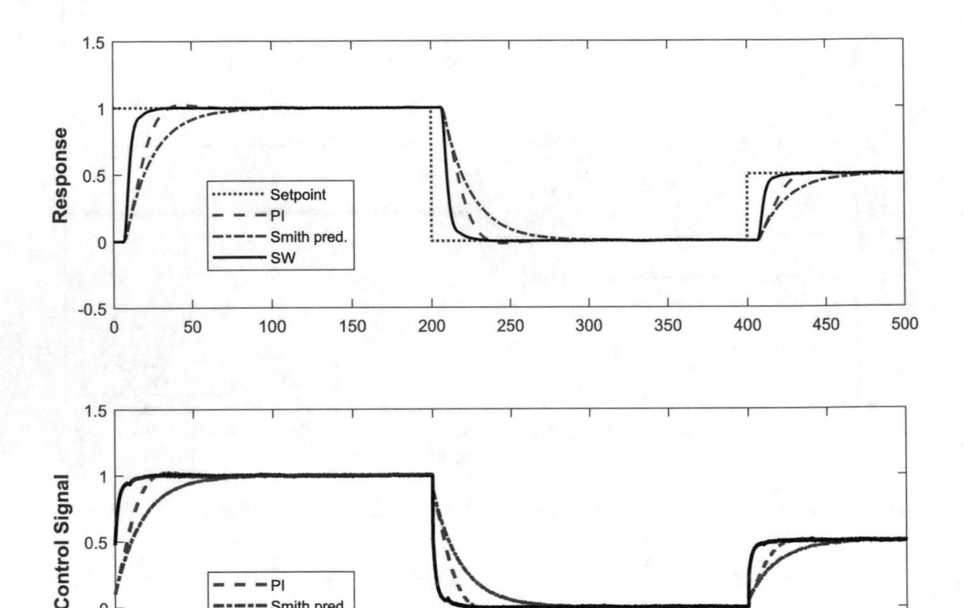

Fig. 3.14 Third order system setpoint tracking

3.4.4 Fourth Order System

Figure 3.15 shows the simulation results for the fourth order system while Fig. 3.16 shows the zoomed-in of regions A, B, C and D of the plots in Fig. 3.15. Numerical results of the two figures are given in Table 3.5. Here, results followed a similar pattern to the three lower order systems considered earlier. The rise time of the plant with proposed SW approach is around 9.5 s. This makes it more than two (2) times faster than that with PI controller and almost than five (5) times that of Smith predictor. The SW controller settled at around 24 s. This is less than half the settling time of PI and around one-fourth of the Smith predictor. This shows that the proposed approach settled much faster than the other controllers. The overshoots produced by

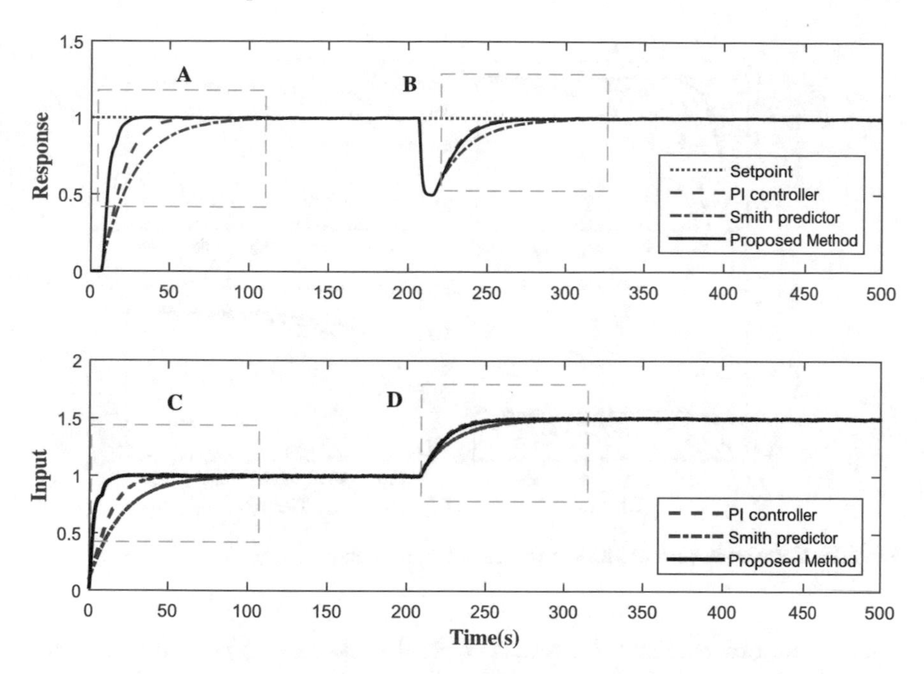

Fig. 3.15 Fourth order system with load disturbance and noise

Table 3.5 Control performance of fourth order system

Controller	System's performance			
	Rise time (s)	Settling time (s)	Overshoot (%)	IAE
SW	9.4962	24.0004	0.6604	306.4781
PI	25.4462	49.8581	0.6671	383.9040
Smith Predictor	45.5658	86.1282	0.5114	511.8992

all control approaches is in the range 0.51–0.67%. This shows that all controllers performed well for this plant. The proposed SW approach also produced the least IAE of around 306.50 as against 383.90 and 511.90 of PI and Smith predictor respectively. The recovery from disturbance effect of the proposed SW method for the fourth order system also followed a similar pattern to the earlier systems considered. The only exception here is that the PI controller sees improvement in its disturbance regulation ability. This is noticeable from its recovery without overshoot as observed from both figures.

Observing the various control signals, the proposed approach produces more aggressive signal than other controllers during step change (see regions A and C of Fig. 3.16). At the point of disturbance, a control signal similar to that of PI and more aggressive to that of Smith predictor is produced by the SW controller (see region D).

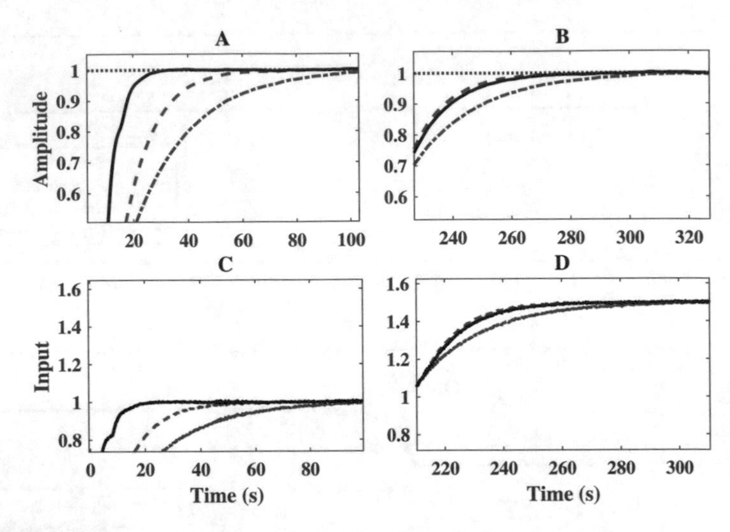

Fig. 3.16 Zoomed-in view of Fig. 3.15 for regions A, B, C and D

A careful observation of the results of Fig. 3.17 reveals the performance of SW controller for tracking changing setpoint on fourth order system. Similar trends from the other plants can clearly be observed i.e. the proposed SW approach outperformed the PI and Smith predictor in terms of tracking ability. This can further be corroborated when various controller actions are observed in the same figure, where it is seen that the proposed approach is more aggressive than the others.

3.4.5 WH-HIL Simulation of First Order System to Real-Time Variable Delay

To verify the viability of the proposed method for application in the real-time wireless environment, an experiment is conducted using WH-HIL simulator developed in our laboratory. The WH-HIL simulator enables the real-time delay to be used for online simulations. It is expected that the design should work in real-life as it works with the simulator.

For the WH-HILS, first order system of Eq. (2.34) is used. Figure 3.18 shows the responses of the system to the real-time delay. Within the figure, the controller actions are also shown Compared to that of PI controller, the response of the proposed method with real-time variable delay performed remarkably better. It is clearly seen from Fig. 3.9 the proposed method has the ability to track changing setpoint without overshoots under real wireless communication conditions.

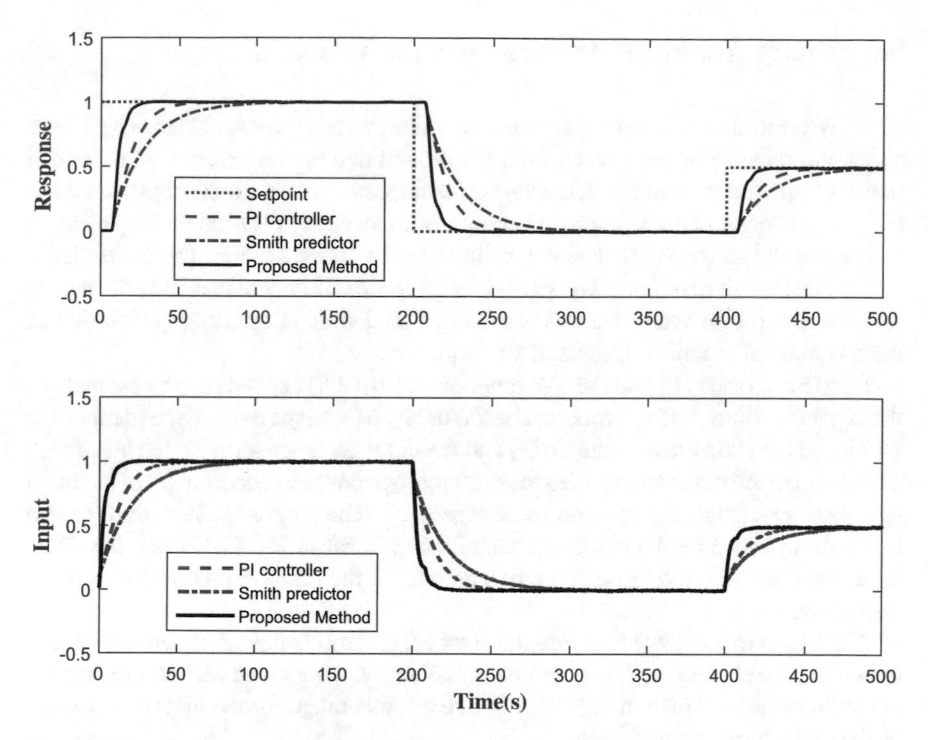

Fig. 3.17 Fourth order system setpoint tracking

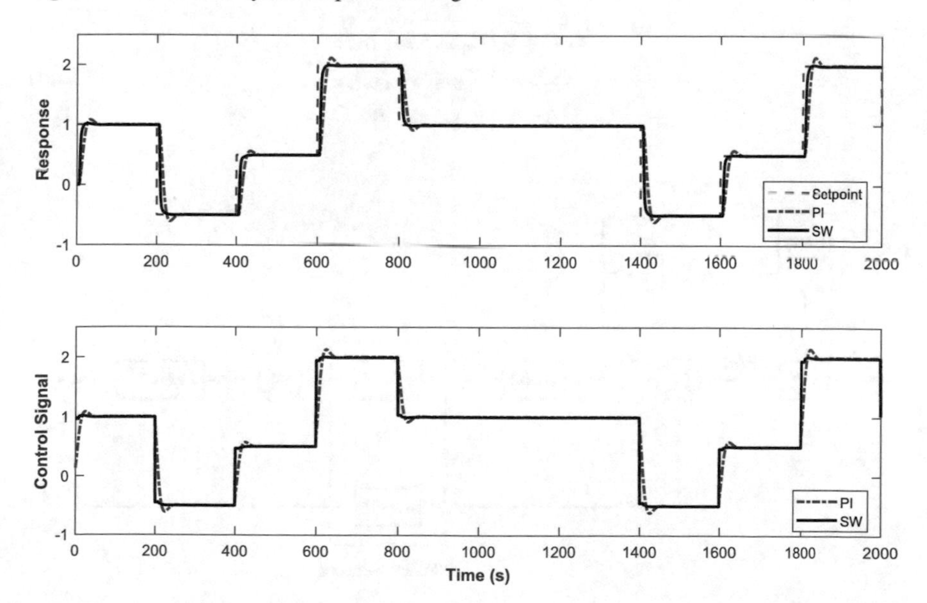

Fig. 3.18 First order system with changing setpoint and experimental real-time variable delay

3.5 Fuzzy Adaptive SW Structure for WHNCS

The SW controller is a very good control strategy for systems characterized by a stochastic delay. However, when used for second and higher order systems, and if there is high variation in the delay, its performance degrades. In this case, although the general performance is improved compared to let us say PID and Smith predictor, but there is the presence of overshoot in the response. Thus, adapting this controller to the stochastic nature of the delay through fuzzy logic will ensure that its performance remains acceptable even with this variation. The details of the fuzzy adaptation of the SW controller will be discussed subsequently.

It has been observed that the SW function of Eq. (3.5) depends on the estimate of the plant deadtime and network stochastic delay. Most especially, dependent terms within the function are the gain $G_r(s)$ and the delay estimate term $e^{-s\tau}$. Thus, fuzzy adaptation mechanism will be used to adjust these parameters accordingly to maintain smooth setpoint tracking and good load regulation. The proposed adaptation scheme is shown in Fig. 3.19. Under the scheme, the K term in Eq. (3.17) and the delay term τ will be adjusted through the two outputs of the fuzzy regulator ΔK and $\Delta\tau$ respectively.

The inputs to the fuzzy tuner are the error (e) and its change (Δe). During on-line operation, the change in $f_r(s)$ parameters ΔK and $\Delta\tau$ are tuned each sampling time using fuzzy tuner shown in Fig. 3.19. The respective ranges of the input and outputs of the fuzzy tuner are as follows:

$$\begin{aligned} e, \Delta e &\in [e_{min}, e_{max}], \\ \Delta K &\in [K_{min}, K_{max}], \\ \Delta\tau &\in [\tau_{min}, \tau_{max}]. \end{aligned} \tag{3.26}$$

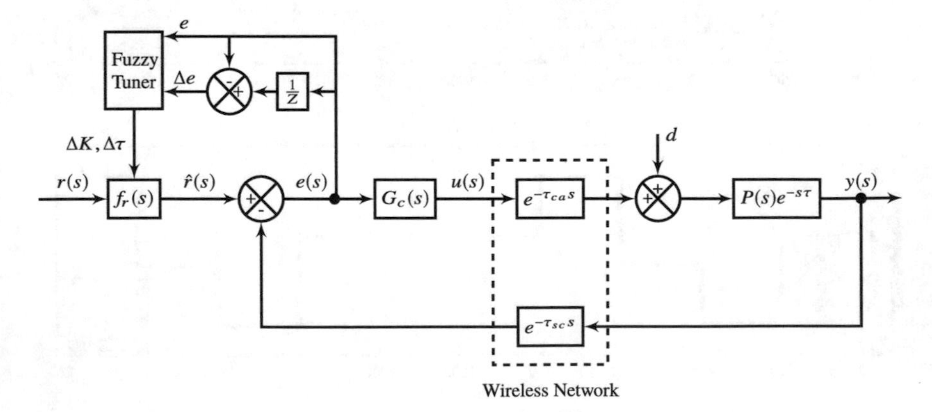

Fig. 3.19 Fuzzy Adaptive SW structure with WirelessHART network

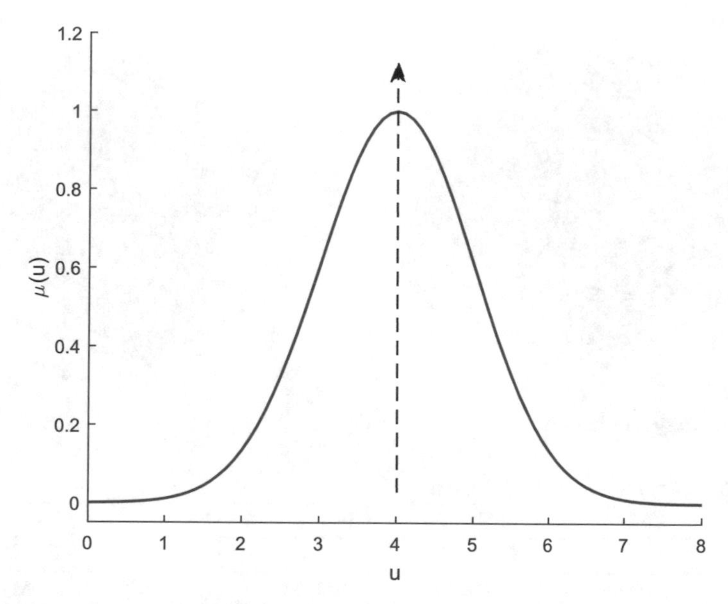

Fig. 3.20 Gaussain membership function

where the subscripts min and max refers to minimum and maximum values of the parameters e, K and τ. The range is selected based on the information obtained from the variation of the WirelessHART network delay.

In this proposed fuzzy adaptation approach, the control rules are developed with e and Δe as a premise and ΔK and $\Delta\tau$ as consequent of each rule. An example of the tuning rule is given as "IF is NB and is NB THEN is NVB and is Z". For smooth adaptation, five and nine Gaussian membership functions are selected for input and output variables respectively. Each of the membership function takes the form shown in Fig. 3.20. The equation of the membership function is given as

$$\mu(x; \sigma, c) = e^{-\dfrac{(x-c)^2}{2\sigma^2}} \tag{3.27}$$

where c and σ are constants defining the shape of the membership function. For the function shown in the figure these constants are chosen as 4 and 1 respectively.

The description of the input membership functions is NB (Negative Big), NS (Negative Small), Z (Zero), PS (Positive Small) and PB (Positive Big). These membership functions are used for both input variables e and Δe. On the other hand, the membership functions of the output variables ΔK and $\Delta\tau$ are NVB (Negative Very Big), NB (Negative Big), NM (Negative Medium), NS (Negative Small), Z (Zero), PS (Positive Small), PM (Positive Medium), PB (Positive Big), PVB (Positive Very Big). Similarly, the linguistic descriptions for the output membership functions are Z (Zero), VS (Very Small), S (Small), SM (Small-Medium), M (Medium), SB (Small Big), MB (Medium Big), B (Big) and VB (Very Big).

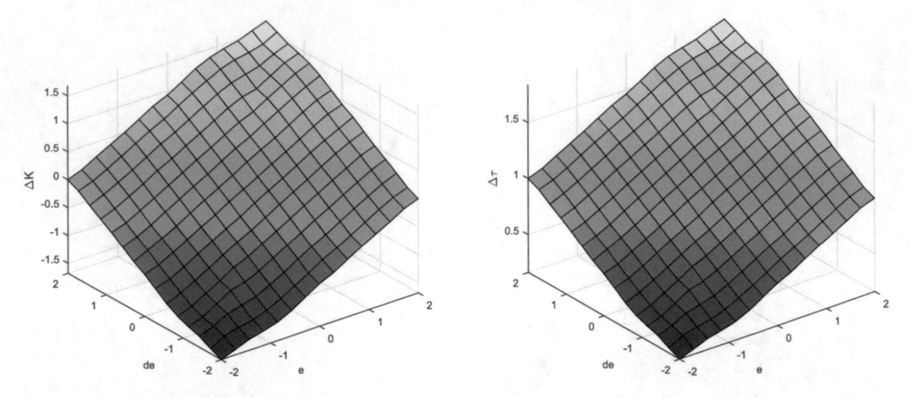

Fig. 3.21 Fuzzy rule surface

Table 3.6 Fuzzy rules for SW adaptation

Δ E *E*	NB	NS	Z	PS	PB
NB	(NVB, Z)	(NB, VS)	(NM, S)	(NS, SM)	(Z, M)
NS	(NB, VS)	(NM, S)	(NS, SM)	(Z, M)	(PS, SB)
Z	(NM, S)	(NS, SM)	(Z, M)	(PS, SB)	(PM, MB)
PS	(NS, SM)	(Z, M)	(PS, SB)	(PM, MB)	(PB, B)
PB	(Z, M)	(PS, SB)	(PM, MB)	(PB, B)	(PVB, VB)

The 25 fuzzy rules are given in Table 3.1. As seen from the table, the first argument of the output represents ΔK while the second argument represents $\Delta\tau$ i.e $(\Delta K, \Delta\tau)$. The respective rule surfaces for the two outputs based on Table 3.1 is given in Fig. 3.21 (Table 3.6).

Fuzzification is achieved using the intersection minimum operation given as follows

$$\mu_{A\cap B}(x, y) = min(\mu_A(x, y), \mu_B(x, y)) \tag{3.28}$$

where A and B are input fuzzy sets (e and Δe). The values of these inputs are calculated at each sampling time as follows

$$e(t) = r(t) - y(t) \tag{3.29}$$

$$\Delta e = \Delta e(t) - \Delta e(t-1) \tag{3.30}$$

For defuzzification, the commonly used centroid method is selected for finding the crisp value of the output. The centroid method is given as:

$$\mu_o = \frac{\sum_{i=1}^{R} c_i \mu_i}{\sum_{i=1}^{R} \mu_i} \tag{3.31}$$

where, μ_o is the fuzzy output, c_i is the centre of the membership function of the consequent ith rule, μ_i is the membership value of the premise ith rule and R is the total number of fuzzy rules.

3.6 Performance of Proposed Fuzzy Adaptive SW Controller

To verify the effectiveness of the fuzzy adaptation of the SW technique, a different delay profile of Fig. 3.22 is considered. The respective maximum, minimum and mean values of the upstream delay as 2.084, 1.2140, and 1.5734 s. On the other hand, the downstream delay is kept constant at 1.28 s. As stated earlier, the performance of the proposed Fuzzy Adaptive SW (FASW) strategy is compared to those of PI and Fixed SW in the time domain. Similar plants as considered for SW are used for simulation.

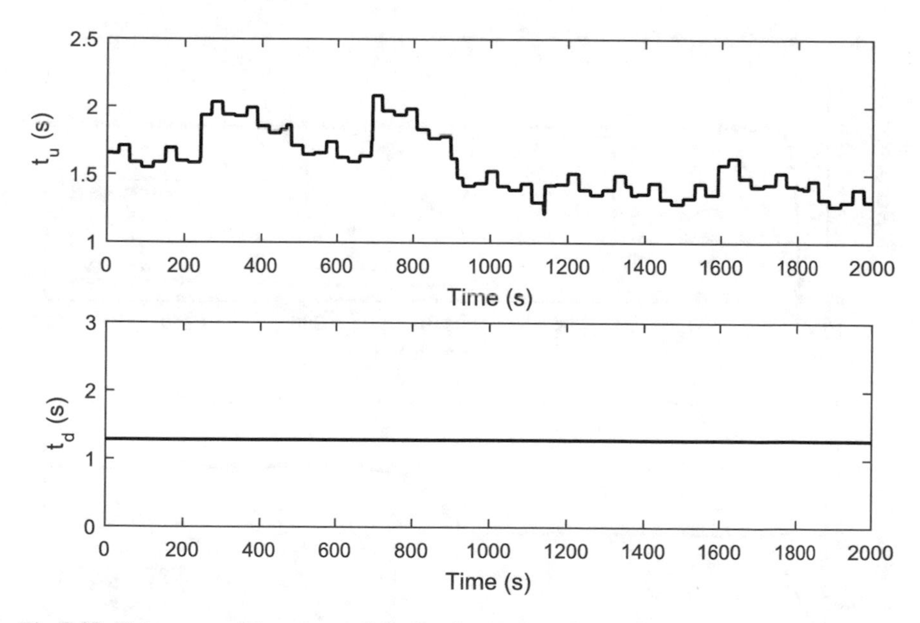

Fig. 3.22 Upstream and downstream delay for fuzzy adaptation

3.6.1 First Order System

The comparison of closed loop response of the first order system for setpoint tracking and disturbance rejection with various controllers is shown in Fig. 3.23 while numerical results are given in Table 3.7. From the figure, it is clearly seen that FASW and SW configurations recorded very close performance in terms of overshoot, IAE and both rise and settling times. For example, the overshoots of both FASW and SW are less than 1% with respective values 0.0727 and 0.2306%. The rise and settling times of both controllers are around 2.7 s and 16 s while their IAE is approximately 33.7. On the other hand, the PI controller recorded the highest overshoot, rise time, settling time and IAE with respective values of 3.4722%, 25.13 s, 75.59 s and 46.32. Furthermore, both controllers recovered from disturbance effect without overshoot as against PI with little overshoot. Observing the control signal of the compared controllers, it can be seen that both SW and FASW starts at around 1 compared to that of PI starting around 0.1. This shows how aggressive the SW controllers can be. Therefore, the FASW control for this plant has little improvement on the SW. Meanwhile, the SW controllers as observed have both significantly improved the performance of the PI.

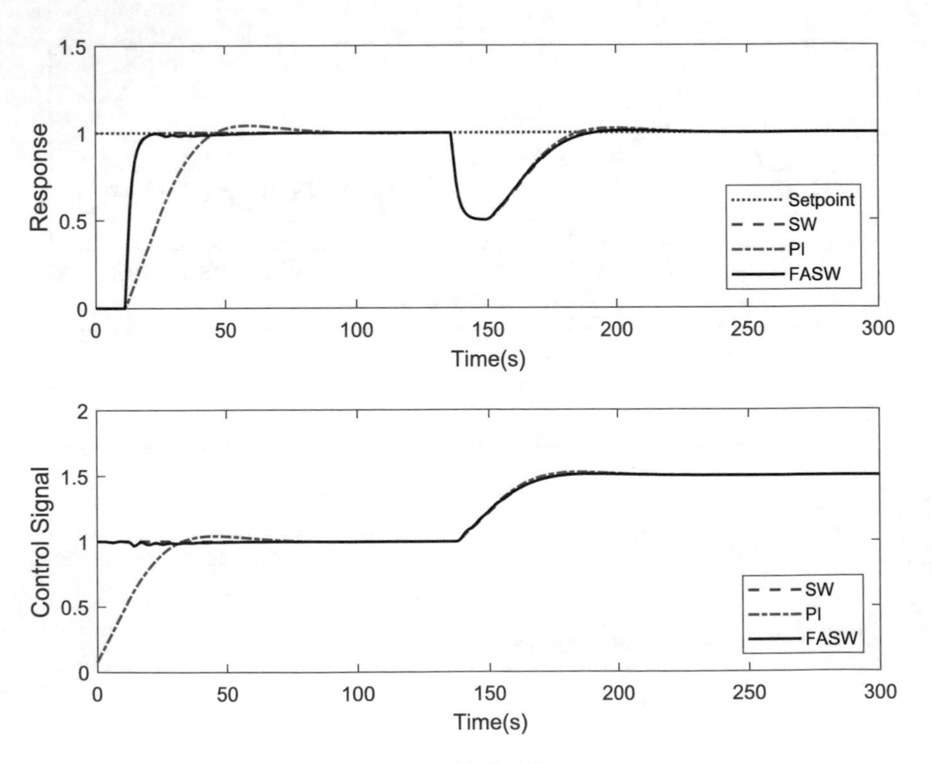

Fig. 3.23 Step response of first order system with disturbance

Table 3.7 Tracking performance of first order system

Controller	System's performance			
	Rise time (s)	Settling time (s)	Overshoot (%)	IAE
FASW	2.7306	16.0417	0.0727	33.7207
SW	2.7790	15.6053	0.2306	33.7327
PI	25.1320	75.5902	3.4722	46.3247

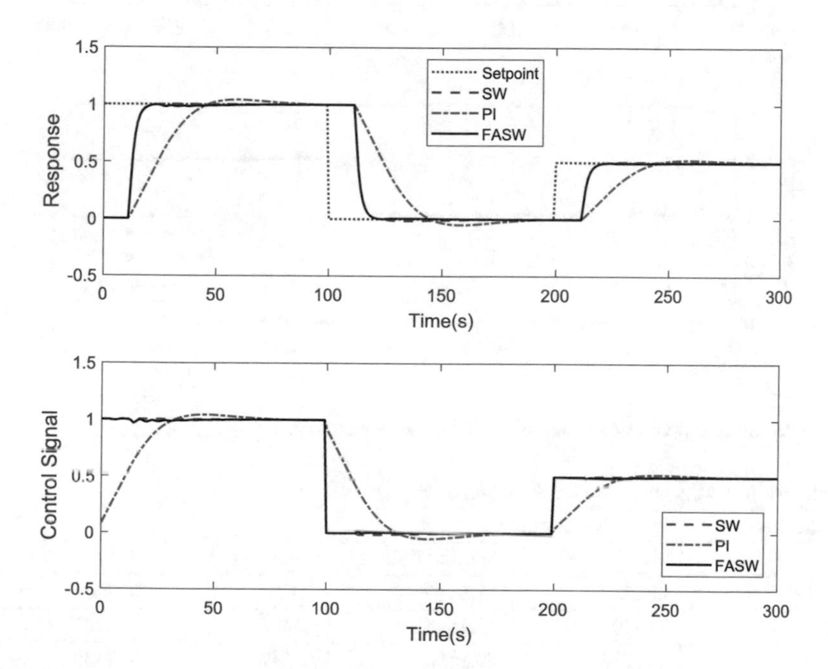

Fig. 3.24 Variable setpoint tracking of various controller for first order system

The comparison of variable setpoint tracking ability with various controllers is shown in Fig. 3.24. From the responses, It can be observed that the tracking performance of SW and FASW are similar and followed the same pattern as to the step response. Furthermore, the two responses are also better than that PI in terms of overshoot and undershoot during setpoint change. The control signals of both SW and FASW also followed a similar trend to those obtained for step response.

3.6.2 Second Order System

The comparison of closed loop response of the second order system for setpoint tracking and disturbance rejection with various controllers is shown in Fig. 3.25 while numerical results are given in Table 3.8. From both the figure and table, it is clearly

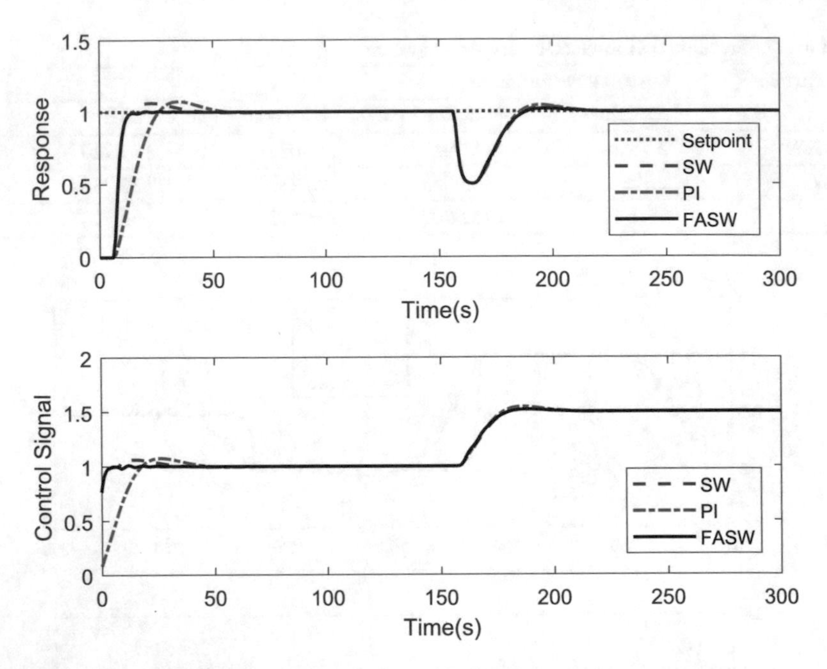

Fig. 3.25 Step response of second order system with disturbance

Table 3.8 Tracking performance of second order system

Controller	System's performance			
	Rise time (s)	Settling time (s)	Overshoot (%)	IAE
FASW	3.8653	11.8789	0.0286	28.4034
SW	4.1330	37.1544	6.1605	30.0163
PI	14.1246	49.2130	7.3542	36.7180

seen that the fuzzy adaptive SW (FASW) configuration achieved the best tracking and disturbance rejection performance with least overshoot of 0.0286% compared to the 6.1605% and 7.3542% of the SW and PI respectively. The configuration also recorded the shortest rise and settling times. It also recorded the least IAE of 28.40 as against 30.01 and 36.72 of the SW and PI respectively. On the recovery from disturbance effect, the SW and FASW controllers showed better recovery with little overshoot compared to the PI controller. Observing the control signal of the compared controllers, it can be seen that both SW and FASW starts at around 0.8 compared to that of PI starting around 0.1. This is similar to the control signals observed for first order plant and clearly shows how aggressive the SW controllers can be.

The comparison of variable setpoint tracking ability with various controllers is shown in Fig. 3.26. From the responses, the tracking performance of FASW is better than those of SW and PI in terms of overshoot and undershoot during setpoint change. This is in conformity with the step response plots.

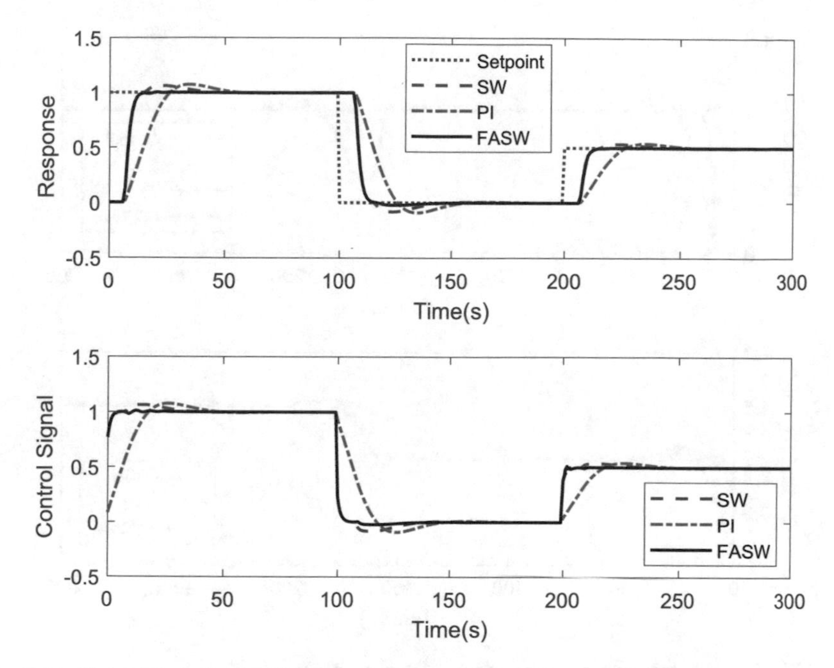

Fig. 3.26 Variable setpoint tracking of various controller for second order system

3.6.3 Third Order System

Similarly, the comparison of closed loop response of the third order system for setpoint tracking and disturbance rejection with various controllers is shown in Fig. 3.27 while numerical results are given in Table 3.9. It is clearly seen from both the figure and table, that the FASW controller achieved the best tracking and disturbance rejection performance with least overshoot of 1.73% as compared to the 9.29% and 8.98% of the SW and PI controllers respectively. In addition, the proposed configuration has the shortest rise time of around 5 s compared to the 7.2 and 13.5 s of the SW and PI controllers. The settling time, IAE and recovery from disturbance also follow a similar pattern to the second order system with FASW outperforming both SW and PI. The SW and FASW as observed from the control signals are more aggressive than the PI controller at the beginning, with both starting at around 0.5 each while PI starts at around 0.1. Hence, the reason for the faster rise and settling times while maintaining good overshoot.

In the same vein, the comparison of variable setpoint tracking ability with various controllers for the third order system is shown in Fig. 3.28. From the responses, just as observed in the second order system, the tracking performance of FASW is better than those of SW and PI in terms of overshoot and undershoot during setpoint change.

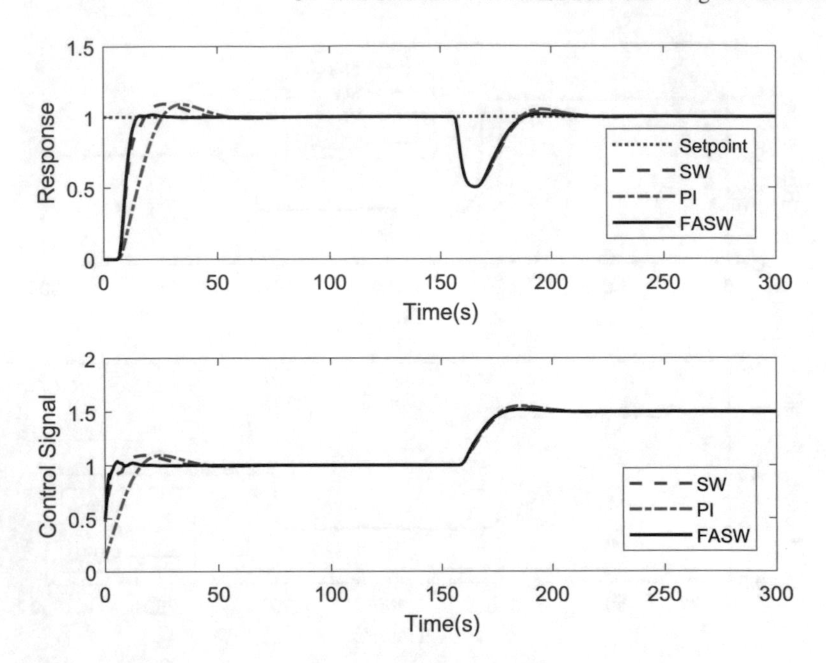

Fig. 3.27 Step response of third order system with disturbance

Table 3.9 Tracking performance of third order system

Controller	System's performance			
	Rise time (s)	Settling time (s)	Overshoot (%)	IAE
FASW	4.9646	14.0577	1.7370	40.6865
SW	7.1831	42.7725	9.2917	44.1787
PI	13.5466	49.6810	8.9848	50.7877

3.6.4 Fourth Order System

Similarly, the comparison of closed loop response of the fourth order system for set-point tracking and disturbance rejection with various controllers is shown in Fig. 3.29 while numerical results are given in Table 3.10. It is clearly seen from both the figure and table, that the FASW controller achieved the best tracking and disturbance rejection performance with least overshoot of 2.32% as compared to the 11.44 and 21.05% of the PI and SW controllers respectively. In addition, the proposed configuration has the shortest rise time of around 2 s compared to the around 9.2 and 18 s of the SW and PI controllers. The settling time, IAE and recovery from disturbance also follow a similar pattern to the second and third order systems. The FASW as observed from the control signals is more aggressive than the both SW and PI controllers at the beginning starting at around 1.30 while PI and SW start at around 0.1 and 0 respectively.

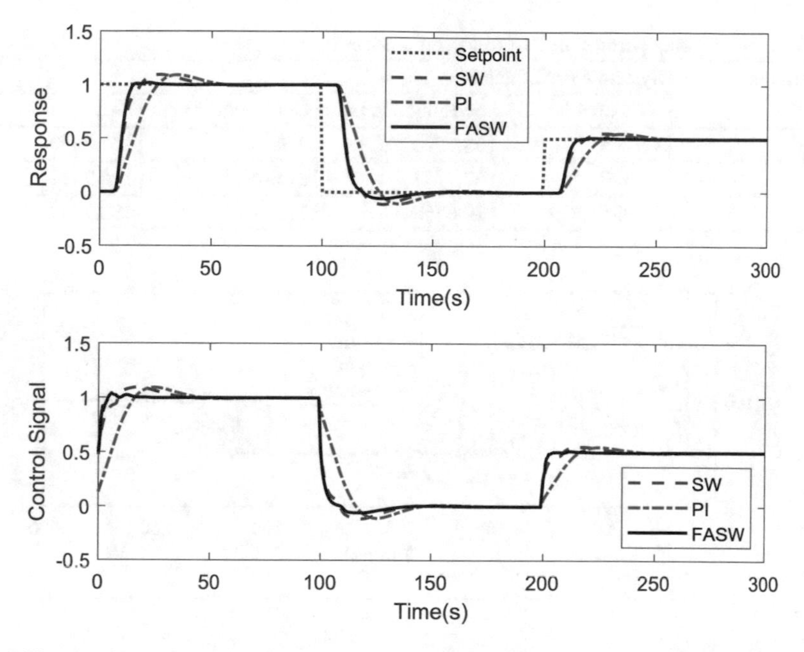

Fig. 3.28 Variable setpoint tracking of various controller for third order system

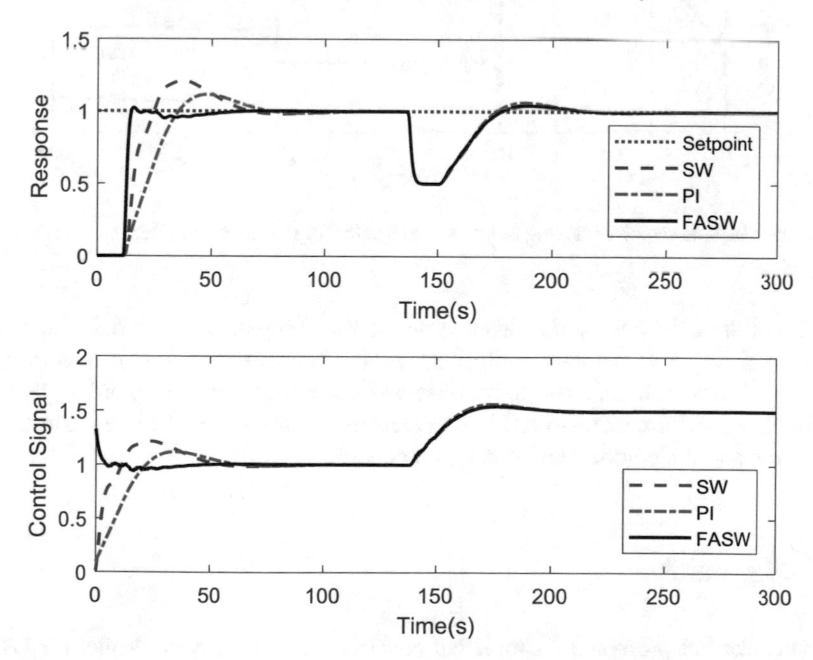

Fig. 3.29 Step response of fourth order system with disturbance

Table 3.10 Tracking performance of fourth order system

Controller	System's performance			
	Rise time (s)	Settling time (s)	Overshoot (%)	IAE
FASW	1.9496	48.7592	2.3211	137.8393
SW	9.2500	62.7777	21.0451	175.7891
PI	17.9941	70.0769	11.4407	192.7799

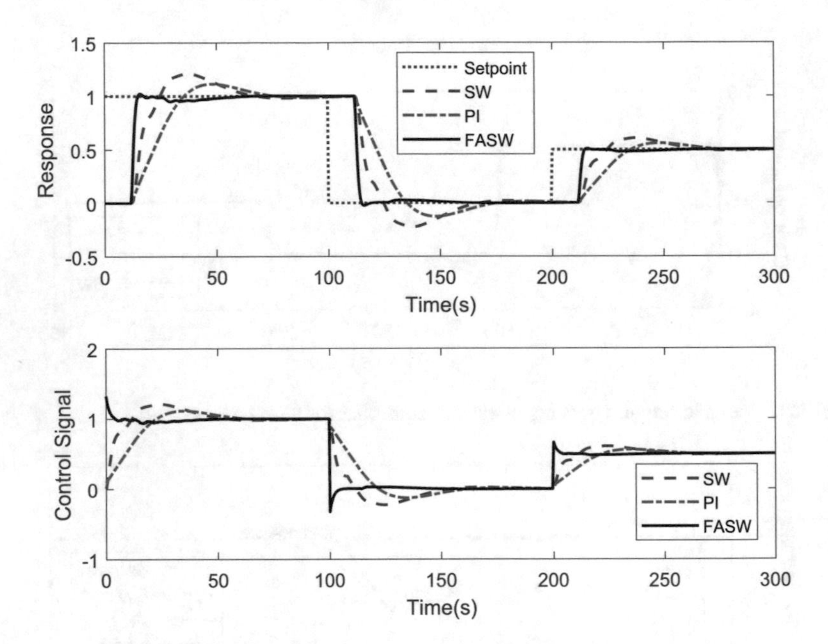

Fig. 3.30 Variable setpoint tracking of various controller for fourth order system

In a similar fashion to the other systems, the comparison of variable setpoint tracking ability with various controllers for the fourth order system is shown in Fig. 3.30. From the responses, just as observed in the second and third order systems, the tracking performance of FASW is outperformed those of SW and PI in terms of overshoot and undershoot during setpoint change.

3.7 Summary

This chapter has presented a simple but powerful PI based SW controller for WirelessHART networked Control systems. This method can be extended to other wireless networked systems and other systems characterised by stochastic network delay. This is evident from the fact that the SW control allow for the use of PID controller in an environment characterised by large delays thereby overcoming the PID's wide gain limitation. The fuzzy adaptation of the design as shown in the later part of the chapter also ensures that even in the presence of wider network variations, the PID continues to give satisfactory response as shown from results presented. Through this design original PID structure can be maintained while a plug and play setpoint weighting function can be introduced without making significant changes to the systems settings.

References

1. Tan, K.K., Tang, K.Z., Su, Y., Lee, T.H., Hang, C.C.: Deadtime compensation via setpoint variation. J. Process Control. **20**(7), 848–859 (2010)
2. Åström, K.J., Hägglund, T.: Advanced PID Control. ISA-The Instrumentation Systems, and Automation Society, Research Triangle Park, NC (2006)
3. Alfaro, V.M., Vilanova, R., Arrieta, O.: Considerations on set-point weight choice for 2-DoF PID controllers. IFAC Proc. Vol. **42**(11), 721–726 (2009)
4. Bianchi, F.D., Mantz, R.J., Christiansen, C.F.: Multivariable PID control with set-point weighting via BMI optimisation. Automatica **44**(2), 472–478 (2008)
5. Mudi, R.K., Dey, C.: Performance improvement of PI controllers through dynamic set-point weighting. ISA Trans. **50**(2), 220–230 (2011)
6. YEŞİL, E., Güzelkaya, M., Eksin, I., Tekin, O.A.: Online tuning of set-point regulator with a blending mechanism using PI controller. Turk. J. Electr. Eng. Comput. Sci. **16**(2), 143–157 (2008)
7. Yeşil, E., Kaya, M., Siradag, S.: Fuzzy forecast combiner design for fast fashion demand forecasting. In: 2012 International Symposium on Innovations in Intelligent Systems and Applications, Trabzon, Turkey, (2–4 July 2012)
8. Visioli, A.: Fuzzy logic based set-point weight tuning of PID controllers. IEEE Trans. Syst. Man Cybern. Part A-Syst. Hum. **29**(6), 587–592 (1999)
9. Visioli, A.: Practical PID Control. Springer Science & Business Media, Berlin (2006)
10. Mitra, P., Dey, C., Mudi, R. K.: Dynamic set-point weighted fuzzy PID controller. In: 2013 International Symposium on Computational and Business Intelligence, New Delhi, India (24–26 August 2013)
11. Paul, P.M., Dey, C., Mudi, R.K.: An online dynamic set point weighting scheme for PID controller. In: Proceedings of the 2014 IEEE Students' Technology Symposium, Kharagpur, India, 28 Feb. âĂ2 March (2014)
12. Hassan, S.M., Ibrahim, R., Saad, N., Asirvadam, V.S., Chung, T.D.: Setpoint weighted WirelessHART networked control of process plant. In: 2016 IEEE International Instrumentation and Measurement Technology Conference Proceedings, Taipei, Taiwan (23–26 May 2016)
13. Ingimundarson, A., Hägglund, T.: Robust tuning procedures of dead-time compensating controllers. Control Eng. Pract. **9**(11), 1195–1208 (2001)
14. Smith, C.A., Corripio, A.B.: Principles and practice of automatic process control. Wiley, New York (1985)

Part II
Hybrid Metaheuristics Algorithm Based Controllers for WHNCS

Chapter 4
Hybrid APSO–Spiral Dynamic Algorithm

4.1 Introduction

The accelerated particle swarm optimisation (APSO) is an improved variant of the PSO algorithm that guarantees convergence through the use of only global best to update both velocity and position of particles. However, like its predecessor, the APSO is also prone to being trapped in local minima. Therefore, this chapter proposes two hybrid algorithms synergizing the social ability of the APSO and the exploitative ability of both spiral dynamic algorithm (SDA) and Adaptive SDA (ASDA). The exploration phase of the proposed algorithms APSO-SDA and APSO-ASDA, will be achieved through the APSO algorithm. The exploration phase solutions of the APSO are then fed to the SDA and ASDA to achieve the exploitation phase.

The proposed algorithms will be evaluated with benchmark function and will be used to tune a filtered predictive proportional-integral (FPPI) controller for WirelessHART networked control systems (WHNCS). The Friedman's rank test will be used to show that the proposed APSO-SDA and APSO-ASDA have the ability to outperform their constituent algorithms. Time domain analysis of the FPPI controller will also be conducted to corraborate the Friedman's rank test. This will be done through evaluation of settling times and overshoots.

The chapter will begin by reviewing briefly some related heuristic algorithms. Then, brief description of the constituent algorithms namely APSO and SDA will be given in successive order. This will be followed by small discussion on the adaptation of SDA. The next section of the chapter will focus on the proposed algorithm which will be followed by validation with established benchmark functions. In the later part of the chapter, the proposed algorithms will be used to tune FPPI as earlier mentioned while results obtained using these algorithms will be compared to those of the constituent algorithms. Finally, a summary will be provided.

S. M. Hassan et al., *Hybrid PID Based Predictive Control Strategies for WirelessHART Networked Control Systems*, Studies in Systems, Decision and Control 293,
https://doi.org/10.1007/978-3-030-47737-0_4

4.2 Brief Overview of Related Heuristics Algorithms

Metaheuristic algorithms combine both simple local search and higher-level search strategies to seek for best solutions of optimisation problems. In such algorithms, an iteratively generated process directs gradient free subordinate heuristics by intelligently combining several explorations and exploitation concepts within the search space [1]. These algorithms are often inspired by natural phenomena such as birds flocking [2], natural selection [3], bacterial foraging [5], path seeking behaviour of ants [6], natural spiral phenomena [7, 8], whale optimisation [11, 12], etc. Key advantage of these algorithms is that high quality solutions can be obtained within short time without the need for gradient of the optimisation problem. This indicates that the differentiability of the search space is not required by such algorithms. Furthermore, metaheuristics can be used as an efficient tools for solving broad range of intractable problems.

The APSO is a simplified form of the PSO, which is guaranteed to have the same or even better optimisation ability. In the APSO, the velocity update equation uses only global best as against using both global and local best in the standard PSO algorithm. This speeds up the convergence of the optimisation process. In this simplified version, the position update can be written without the velocity term in order to further hasten the convergence. However, despite the advantages of convergence speed, the APSO like its parent algorithm, is still prone to being trapped in local minima. Recently, researchers have either applied or attempted to improve on the original APSO algorithm. For example in [13], the algorithm has been applied to optimise parameters of wire electrical discharge machining (WEDM) to achieve minimum surface roughness, maximum Material Removal Rate (MRR) and minimum kerf width. Authors in [14] have applied the algorithm to achieve maximum power point (MPPT) in a partially shaded PV system. A dynamic spectrum sensing algorithm based on APSO is proposed in [15]. In that work, it was found that the proposed method compared to the existing ones, gives more efficient results. Wang et al. [16] proposed an improvement to the APSO algorithm by hybridising it with differential evolution (DE) algorithm. In the paper, instead of using random walk, the mutation operator of DE is used to tune the update of velocity and positions of each particle. In another work by Guedria [17], an improvement to the APSO algorithm through use of individual particle memories to update positions of particles in the swarm has been presented. Results comparison showed that the improved algorithms can perform even better than the compared algorithms.

The spiral dynamic algorithm (SDA) was proposed first for two-dimensional problems by Tamura and Yasuda in [7]. The authors later extended the algorithm for n-dimensional problems in [8]. The algorithm mimics natural spiral phenomena like galaxy, tornadoes and hurricane. The SDA compared to other known metaheuristic algorithms, is simple hence it has short computational time and faster convergence speed. A unique feature of the SDA is that in the initial stage of the algorithm, diversification strategy is employed to coarsely explore wider region in the search space. In the later stage of the algorithm, intensification strategy is employed to intensively and finely search around region of good solution. The strategy of diversification and intensification is a pointer to an effective meta-heuristic optimisation [1, 8]. Most of the improvements to the algorithm are geared towards adaptation and hybridization. In [18] several adaptation schemes for the spiral radius ranging from fuzzy, linear, quadratic and exponential functions have been proposed. These schemes are all based on the fitness values to change dynamically the spiral radius. In a related development, Nasir et al. [19] proposed adaptation of both the spiral radius and spiral angle of rotation using linear function. Majority of the hybridization effort made for the algorithm are with bacterial foraging algorithm (BFA). For instance authors of [1] proposed hybridization of bacteria-chemotaxis with SDA. Here, the exploration is achieved through the BFA while the exploitation phase is handled by the SDA. In other works also by same authors [20, 21], the elimination and dispersal ability of the BFA has been employed to improve the search ability of the SDA. Furthermore, a proposal to update the bacterial swim in the BFA algorithm in a spirally based on SDA has been proposed in [22]. A key drawback of these hybridizations is that they are all based on the BFA which is complex and requires high computational time. It should be noted that to our knowledge, there is no reported case on synergy between SDA and APSO algorithm.

The first objective of this chapter is to propose synergy between the APSO algorithm and both the SDA and adaptive SDA (ASDA). The second aim of the chapter is to demonstrate the effectiveness of this proposal through tuning of a filtered PPI (FPPI) controller in WirelessHART environment. Assessment of the performance of proposed approaches will be based on comparison with predecessor algorithms i.e. APSO, SDA and ASDA.

Hence, subsequently, we will present the hybridization of APSO algorithm with SDA and ASDA. First, brief highlight on the constituent algorithms namely APSO, SDA and ASDA will be given. Then, the development of hybrid APSO-SDA and APSO-ASDA will be presented.

4.3 Accelerated Particle Swarm Optimisation Algorithm

The standard PSO algorithm is based on the swarm behaviour of particles and is governed by two update equations i.e. Eqs. (4.1) and (4.2) that represents the velocity and position of each particle in the swarm. At the initial stage of the algorithm, particles are randomly distributed in the n-dimensional search space. Then the velocity and position are updated with the corresponding values of the previous iterations.

$$v_i(t+1) = v_i(t) + c_1\varphi_1(X_i(t) - x_i(t)) + c_2\varphi_2(X_g(t) - x_i(t)) \tag{4.1}$$

$$x_i(t+1) = x_i(t) + v_i(t+1) \tag{4.2}$$

where, c_1 and c_2 are constants representing learning rates, φ_1 and φ_2 are uniformly distributed random numbers, $v_i(t)$ and $x_i(t)$ are velocity and position of the ith particle and $X_i(t)$ and $X_g(t)$ are local and global best positions of the particle.

As seen in Eqs. (4.1) and (4.2), the algorithm uses current values of both local and global positions to update the velocity. This was done to increase the diversity of solution. However, Yang [3, 4] suggested that since it is possible to employ some randomness to simulate the diversity, then local best is not needed in the velocity update. This is to avoid the particles being trapped in a local minimum. This improved version of the PSO i.e., the Accelerated PSO (APSO), uses only the global best for velocity update, hence accelerating the convergence of the algorithm. The velocity and position update equations of the APSO are given as follows

$$v_i(t+1) = v_i(t) + \rho\epsilon_n + \sigma(X_g(t) - x_i(t)) \tag{4.3}$$

$$x_i(t+1) = x_i(t) + v_i(t+1) \tag{4.4}$$

To further step-up the convergence, Eq. (4.4) can be written in a single step as follows:

$$x_i(t+1) = (1-\sigma)x_i(t) + \sigma X_g(t) + \rho\epsilon_n \tag{4.5}$$

where σ is a constant chosen between 0.1 and 0.7 for most applications, $\epsilon_n \in N(0, 1)$ is a random number and ρ is a randomness parameter.

The randomness parameter ρ can be selected as constant or using decay function such as

$$\rho = \rho_0 e^{-\gamma t}, (0 < \gamma < 1) \tag{4.6}$$

or

$$\rho = \rho_0 \gamma^t, (0 < \gamma < 1) \tag{4.7}$$

where $\rho_0 \approx 0.5 - 1$ is the initial randomness parameter, γ is a control variable and t is the number of iterations such that $t \in [0, t_{max}]$. In this work Eq. (4.6) is used for APSO and in both hybrid algorithms. The flowchart for the implementation of both PSO and APSO algorithms is shown in Fig. 4.1.

4.4 Spiral Dynamic Algorithm

The SDA is developed from the inspiration of natural spiral phenomena. Spiral phenomena occur in nature in the form of hurricane, spiral galaxies, whirling current, nautilus shells and low pressure fronts. The spiral dynamic algorithm was proposed by Tamura in [7, 8]. In the algorithm, the spiral model is the main component that defines the characteristics and shape of the spiral. The spiral model depends lagely on two parameters i.e. spiral radius r and spiral rotation angle θ. Theses parameters determine the convergence speed and accuracy of the algorithm.

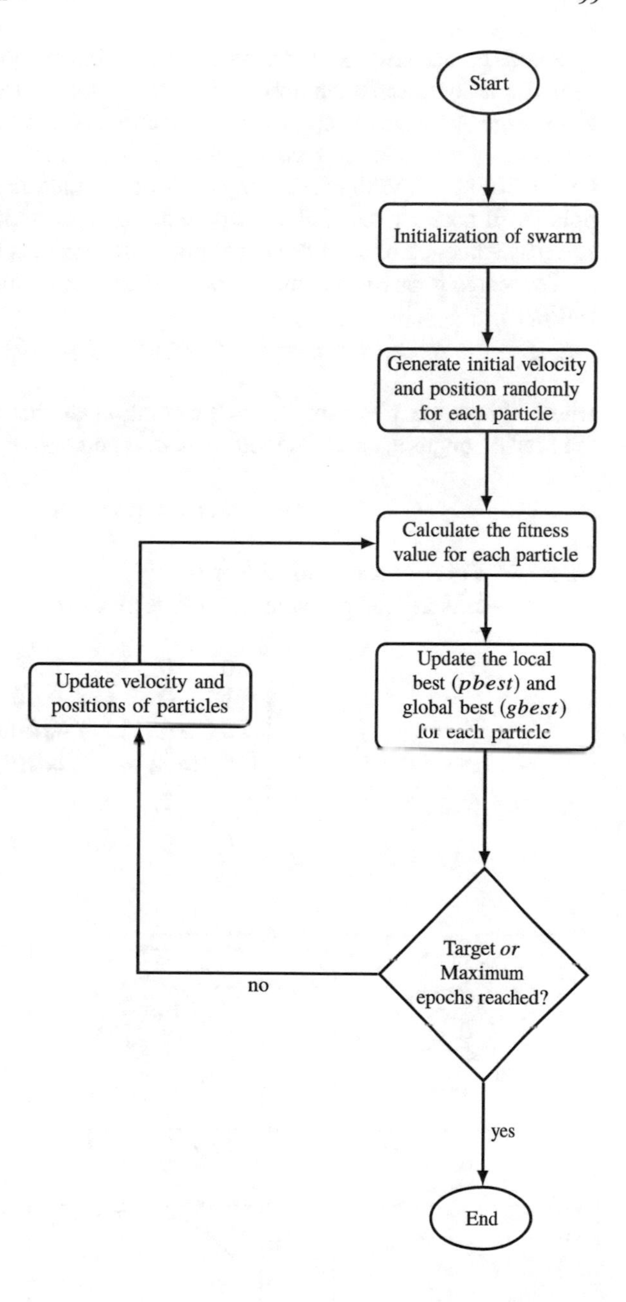

Fig. 4.1 Flowchart for implementation of PSO and APSO algorithms

Search points movement in SDA from the initial position at the outermost layer is geared spirally towards the optimal position at the centre. This is achieved initially using large step-size to achieve diversification. At the later stage, intensification is achieved by dynamically reducing the step-size of the movement at the innermost layer of the spiral. With the strategy of diversification and intensification, the search points will possibly reach global optimum on the search space. The diversification and intensification of a single search point movement is depicted in Fig. 4.2.

The search point movement is modelled using n – dimensional spiral equation as follows

$$x_i(k+1) = S_n(r,\theta)x_i(k) - \big(S_n(r,\theta) - I_n\big)x^* \tag{4.8}$$

where x^* is the centre of spiral, k is the iteration number, r is the rotation radius, θ is the rotation angle, I_n is $n \times n$ identity matrix and $S_n(r,\theta)$ is a stable matrix given as

$$S_n(r,\theta) = r R^{(n)}(\theta) \tag{4.9}$$

where $R^n(\theta)$ is $n \times n$ rotational matrix.

For $n-$dimensional problem, $R^{(n)}(\theta)$ is given as

$$R^{(n)}(\theta) = \begin{bmatrix} 1 & 0 & 0 & \dots & 0 & 0 \\ 0 & 1 & 0 & \dots & 0 & 0 \\ 0 & 0 & cos(\theta_{i,j}) & \dots & -sin(\theta_{i,j}) & 0 \\ 0 & 0 & sin(\theta_{i,j}) & \dots & cos(\theta_{i,j}) & 0 \\ \vdots & \vdots & \vdots & \ddots & \vdots & \vdots \\ 0 & 0 & 0 & \dots & 0 & 1 \end{bmatrix} \tag{4.10}$$

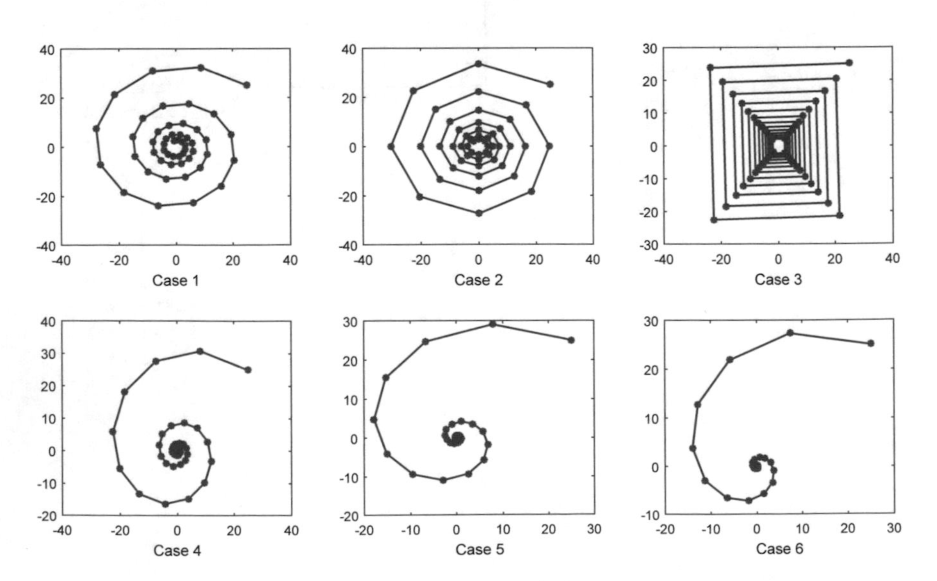

Fig. 4.2 Two-dimensional spiral model: Effect of diversification and intensification in SDA

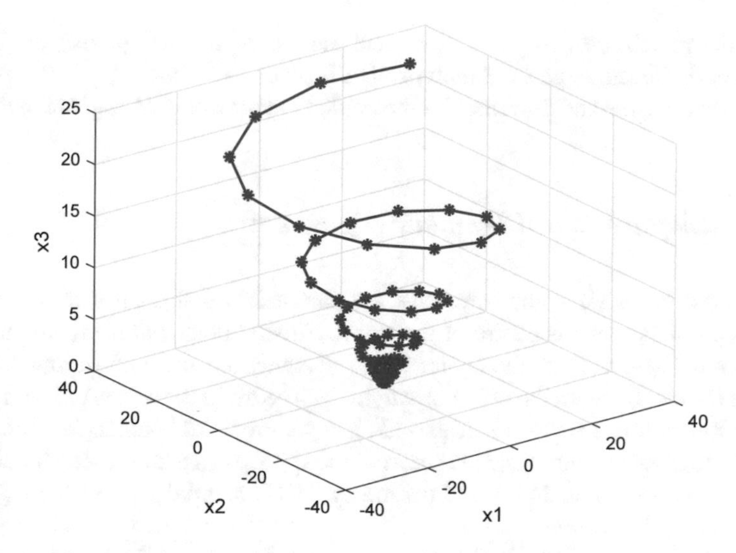

Fig. 4.3 Three-dimensional spiral model

For example, for a two-dimensional problem, the $R^{(2)}(\theta)$ is given as

$$R^{(2)}(\theta) = \begin{bmatrix} cos(\theta) & -sin(\theta) \\ sin(\theta) & cos(\theta) \end{bmatrix} \tag{4.11}$$

The spiral model shown in Fig. 4.5 at $x_0 = (25, 25)$ is for two dimensional problem with $r = 0.95$ and $\theta = \pi/6$. Thus,

$$R^{(2)}(\pi/6) = \begin{bmatrix} 0.866 & -0.500 \\ 0.500 & 0.866 \end{bmatrix} \tag{4.12}$$

In a similar fashion, for three-dimensional problem, one of the three possibilities of $R^{(3)}(\theta)$ is given as

$$R^{(3)}(\theta) = \begin{bmatrix} cos(\theta_{1,2}) & -sin(\theta_{1,2}) & 0 \\ sin(\theta_{1,2}) & cos(\theta_{1,2}) & 0 \\ 0 & 0 & 1 \end{bmatrix} \tag{4.13}$$

For the same values of r and θ as the two-dimensional, the $R^{(3)}(\theta)$ is given in Eq. (4.14) while the spiral model is shown in Fig. 4.3. In this figure, the diversification and intensification effect can also be seen.

$$R^{(3)}(\pi/6) = \begin{bmatrix} 0.866 & -0.500 & 0 \\ 0.500 & 0.866 & 0 \\ 0 & 0 & 1 \end{bmatrix} \tag{4.14}$$

For the remaining two possibilities and the corresponding spirals see [8]. The pseudo-code for the implementation of SDA is given as follows:

Furthermore, the and flowchart for the implementation of SDA is given in Fig. 4.4.

4.4.1 Adaptive Spiral Dynamic Algorithm

In the standard SDA, r and θ are the key parameters that control the movement of search points towards centre of the spiral. This is illustrated using six different scenarios in Fig. 4.5. From the figure, Case 1–3 shows the effect of varying the angle θ while Case 4–6 shows the effect of varying r. The different values of r and θ used for this illustration are given in Table 4.1. The figure clearly depicts how choice of these parameters can influence the shape of the spiral, hence how a search point will move. Thus, poor selection of the parameters will potentially make the algorithm skip global optimal point.

In an attempt to adapt r and θ to the change of fitness of the associated search point, Nasir et al. [19] proposed the following formulations for adaptive radius (r_a) and angle (θ_a):

$$r_a = \frac{(r_l - r_u)}{1 + \left(\dfrac{c_r}{|f(x_i(k)) - minf(x(k))|}\right)} + r_u \tag{4.15}$$

$$\theta_a = \frac{(\theta_l - \theta_u)}{1 + \left(\dfrac{c_a}{|f(x_i(k)) - minf(x(k))|}\right)} + \theta_u \tag{4.16}$$

where, c_r and c_a are positive constant representing fitness deviation rate of change, $f(x_i(k))$ is fitness of a given search point, $minf(x(k))$ is the current iteration's global best fitness, r_l and r_u are the lower and upper values of the spiral radius chosen between 0 and 1, θ_l and θ_u are the minimum and maximum values of the spiral trajectory angle chosen between 0 and 2π.

With these adaptations, when a high fitness point in the search space is detected, this leads to large radius and angle that will yield longer step size. This will move the point to a better location. On the other hand, if low fitness point is detected, smaller radius and angle that will eventually yield shorter step size are obtained. Consequently, this point continue around to search for optimal solutions. Thus, apart from the updates of r and θ with Eqs. (4.15) and (4.16), all other steps in the ASDA are similar to those of the SDA.

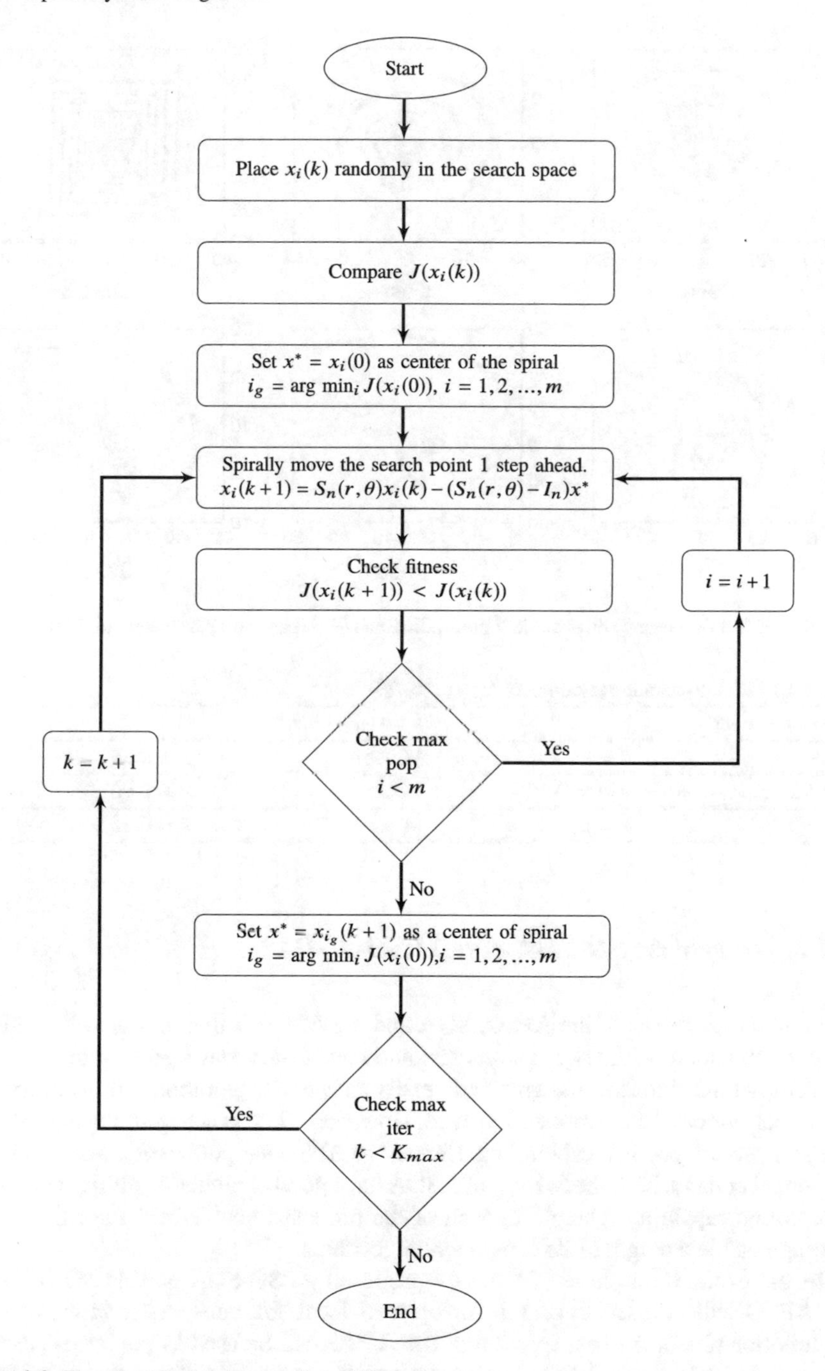

Fig. 4.4 Flowchart for implementation of SDA

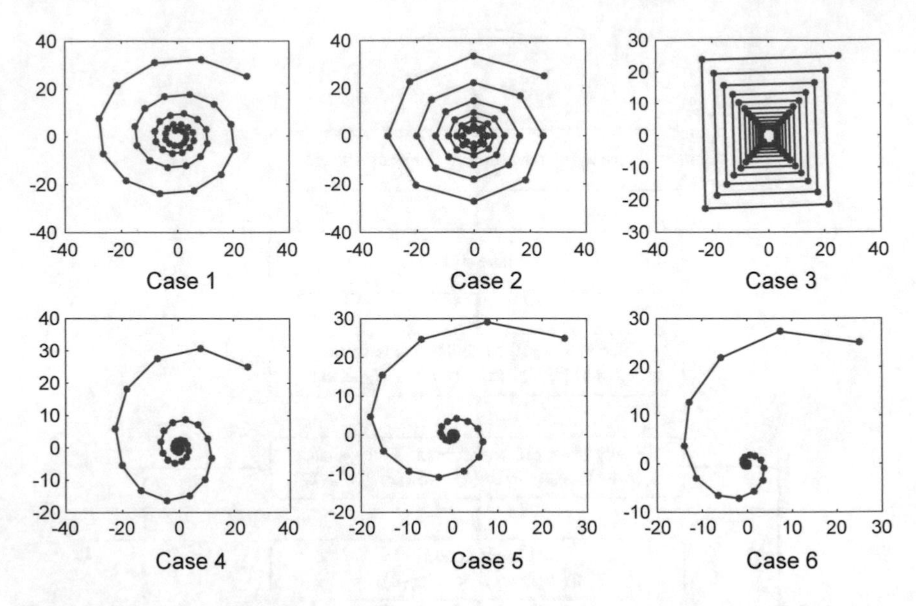

Fig. 4.5 Effect of varying spiral angle θ and radius r in SDA based on parameters of Table 4.1

Table 4.1 SDA Illustration parameters, $x_0 = (25, 25)$

Fixed $r = 0.95$		Fixed $\theta = \pi/6$	
Case 1	$\theta = \pi/6$	Case 4	$r = 0.95$
Case 2	$\theta = \pi/4$	Case 5	$r = 0.90$
Case 3	$\theta = \pi/2$	Case 6	$r = 0.85$

4.4.2 Hybrid APSO-SDA and APSO-ASDA

A common advantage of the APSO, SDA and the ASDA is that they are all simple, easy to implement with fewer parameters and have faster convergence rate. On one hand, the APSO algorithm has greater diversity and good exploration ability due to the social behaviour of the particles involved. However, a key problem of the algorithm is that it gets trapped in local minima. Hence, the APSO has poor exploitation ability. On the other hand, both the SDA, and ASDA have good exploitation ability but poor exploration capability. Therefore, each of the proposed new hybrid algorithm will leverage on the strength of its constituent algorithms.

In the exploration phase of the new proposed APSO-SDA and APSO-ASDA, the APSO will be used to search for optimal local solutions. Afterwards, in the exploitation phase, the best solution of the APSO will be used as the initial search points in the SDA and ASDA to exploit around this good region for global solutions. In other words, the SDA and ASDA will be used to fine-tune the solutions of the APSO. The implementation flowchart of the proposed algorithms is given in Fig. 4.6

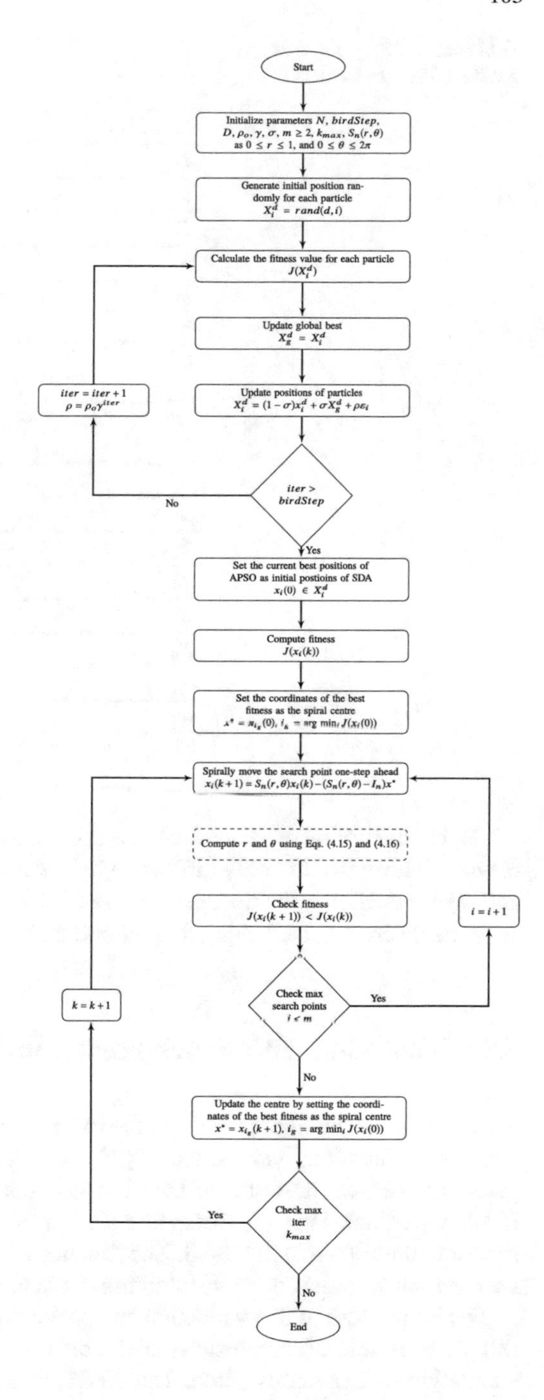

Fig. 4.6 Flowchart for implementation of APSO-SDA and APSO-ASDA

Table 4.2 APSO-SDA and APSO-ASDA parameters

Parameter	Descripton
N	Number of birds
$birdStep$	Number of bird steps
D	Dimension of the problem
J	Fitness funtion
X	Bird position
v	Bird velocity
m	Number of spiral search points
r	Convergence rate or spiral radius
R^n	Rotation matrix of dimension $n \times n$
x^*	Center
x	Rotation point
k_{max}	Max iteration number
I_n	Identity matrix
θ	Rotation angle
S_n	Stable matrix
r_a	Adaptive radius
θ_a	Adaptive angle

while the description of various parameters of the algorithms are given in Table 4.2. It is worth noting that the only difference between the implementation of APSO-SDA and APSO-ASDA is the update of r and θ with (4.15) and (4.16) in the latter. This stage has been indicated using dashed outline in Fig. 4.6.

4.5 Validation with Benchmark Functions

This section will attempt to validate the two proposed algorithms with some selected benchmark functions as reported in [23, 24]. The validation will be done in comparison to APSO, SDA and ASDA. The validation functions considered in include Six-hump camel, Shubert, Booth and Easom functions. The formulations for these functions are shown in Table 4.3. The functions have been selected to reflect various shapes such as valley, many local minima, plate, and steep ridges.

For the purpose of this validation and performance analysis of the proposed algorithms, 3, 10 and 30 dimensions of the benchmark functions for 100 independent simulation runs are considered. The APSO parameters used for all algorithms are defined as $N = 100$, $birdStep = 100$, $\rho_o = 0.5$, $\sigma = 0.7$, $\gamma = 0.7$. Likewise, to ensure fairness, the SDA and ASDA parameters are defined such that the same num-

Table 4.3 Benchmark functions

Function	Formula	Search space	Global minimum
Six-hump camel	$f(x) = \left(4 - 2.1x_1^2 + \frac{x_1^4}{3}\right)x_1^2 + x_1x_2 + (4x_2^2 - 4)x_2^2$	$-3 \leq x_1 \leq 3$, $-2 \leq x_2 \leq 2$	-1.0316
Shubert	$f(x) = \prod_{i=1}^{2}\sum_{j=1}^{5} j cos((j+1)x_i + j)$	$-5.12 \leq x_i \leq 5.12$	-186.7309
Booth	$f(x) = (x_1 + 2x_2 - 7)^2 + (2x_1 + x_2 - 5)^2$	$-10 \leq x_i \leq 10$	0
Easom	$f(x) = -\cos(x_1)\cos(x_2)\exp\left(-(x_1-\pi)^2 - (x_2-\pi)^2\right)$	$-100 \leq x_i \leq 100$	-1

ber of iteration are obtained with the APSO. Thus, their parameters are defined as $m = 100$, $k_{max} = 100$, $r = 0.95$, $\theta = \pi/6$, $r_l = 0.1$, $r_u = 1$, $\theta_l = 0.1$, $\theta_u = 2\pi$, $c_r = 1$ and $c_a = 100$. The selection of these parameters is done through trial and error and mainly guided by the works reported in [7–9, 19]. It should also be noted that the initial particles positions and spiral search points are selected randomly.

The results of 100 simulation run conducted with all the compared algorithms for each of the four benchmark functions is presented in Table 3.3. The results are given in terms of best, worst and in terms of statistical mean and standard deviation (SD). The statistics is to show how consistent and accurate an algorithm is. As indicated in the table, the best of the mean, best, worst and standard deviation is highlighted in bold. For the Six-hump camel function, it can be observed that interms of the Best run, the APSO, APSO-SDA and APSO-ASDA algortihms have achieved the global optimum of -1.0316 for all the 3, 10 and 30 dimensions. For this same function, the APSO-ASDA has best mean for dimensions 3 and 30 while APSO-SDA achieved the best mean for dimension 10. In terms of SD and best of Worst, the same pattern is observed with APSO-ASDA having 2 out of 3 for dimensions 10 and 30 while APSO-SDA has 1 for dimension 3. Results of the Shubert function indicated that the APSO-ASDA algorithm has the best mean and SD for the three dimensions considered. The APSO-SDA has 2 out the 3 best mean for the Booth function while the APSO-ASDA has the one corresponding to 3 dimensions. On the other hand, APSO-ASDA takes 2 of the best SD of the Booth function while the APSO-SDA gets the one corresponding to dimensions 10. The best of mean, best and worst of the optimisation with Easom function have all been dominated by the APSO-ASDA. Although the SDA did not fare well in terms of the mean, best and worst, it won all the SD corresponding to the dimensions considered for the Easom function.

Analysis of the performance of each algorithm with the benchmark function is given in Table 4.4 and Fig. 4.7. The table shows frequencies of best results for each algorithm in terms of mean, best, worst and standard deviation while the figure shows

Table 4.4 Statistical results for 100 simulation runs of benchmark functions

No.	Problem	Dim.		APSO	SDA	ASDA	APSO-SDA	APSO-ASDA
1	Six-hump camel	3	Mean	−1.0257	−0.9034	−0.8794	−1.0270	−1.0273
			Best	−1.0316	−1.0036	−0.9994	−1.0316	−1.0316
			Worst	−0.9720	−0.4896	−0.4675	−1.0029	−0.9911
			SD	0.0091	0.0913	0.1051	0.0059	0.0080
		10	Mean	−1.0246	−0.8998	−0.8823	−1.0273	−1.0266
			Best	−1.0316	−1.0057	−0.9941	−1.0316	−1.0316
			Worst	−0.9151	−0.5595	−0.5020	−0.9780	−0.9885
			SD	0.0175	0.0769	0.1001	0.0074	0.0068
		30	Mean	−1.0254	−0.9016	−0.8775	−1.0259	−1.0282
			Best	−1.0316	−1.0182	−0.9971	−1.0316	−1.0316
			Worst	−0.8708	−0.5949	−0.4380	−0.9310	−0.9786
			SD	0.0169	0.0836	0.1056	0.0122	0.0065
2	Shubert	3	Mean	−183.0735	−139.2888	−135.9428	−183.5993	−183.6704
			Best	−186.7308	−186.3891	−186.5624	−186.7309	−186.7308
			Worst	−123.0031	−55.5230	−47.7837	−123.5655	−123.5309
			SD	11.0900	34.6026	33.5879	11.0035	9.9167
		10	Mean	−182.7526	−133.5925	−135.8872	−182.8331	−183.6709
			Best	−186.7309	−184.8262	−186.1619	−186.7309	−186.7309
			Worst	−106.3559	−45.0788	−51.9855	−121.0531	−110.8807
			SD	12.9030	32.8562	35.0346	12.6377	11.0533
		30	Mean	−182.4027	−140.4786	−135.3453	−182.7113	−183.4608
			Best	−186.7308	−186.7276	−186.4072	−186.7308	−186.7309
			Worst	−121.7422	−45.8797	−56.6269	−116.9337	−121.9951
			SD	12.9474	33.7231	34.6460	13.5574	11.3906

(continued)

Table 4.4 (continued)

No.	Problem	Dim.		APSO	SDA	ASDA	APSO-SDA	APSO-ASDA
3	Booth	3	Mean	0.5990	0.1558	0.0944	0.0172	0.0119
			Best	0.2631	0.0026	6.2242×10^{-4}	4.6759×10^{-7}	7.6309×10^{-5}
			Worst	1.0596	0.6808	0.7325	0.3243	0.0628
			SD	0.1724	0.1408	0.1236	0.0379	0.0158
		10	Mean	0.5657	0.1358	0.0856	0.0174	0.0177
			Best	0.1218	0.0024	0.0022	6.8371×10^{-5}	8.7069×10^{-6}
			Worst	0.9860	0.5940	0.5323	0.1885	0.2181
			SD	0.1607	0.1201	0.0965	0.0280	0.0293
		30	Mean	0.5359	0.1355	0.0865	0.0177	0.0198
			Best	0.1427	0.0099	7.6172×10^{-4}	1.1823×10^{-4}	8.1695×10^{-6}
			Worst	0.9787	0.5301	0.6143	0.3088	0.2263
			SD	0.1630	0.1213	0.1029	0.0366	0.0343
4	Easom	3	Mean	−0.0263	-2.1869×10^{-5}	-2.1032×10^{-5}	−0.0175	−0.9998
			Best	−0.3081	-2.8918×10^{-5}	-2.8902×10^{-5}	−0.2644	−1.0000
			Worst	-8.0105×10^{-5}	-1.3694×10^{-5}	-1.1837×10^{-5}	-8.0392×10^{-5}	−0.9986
			Std. Dev	0.0759	3.6644×10^{-6}	4.1116×10^{-6}	0.0579	2.0853×10^{-4}
		10	Mean	−0.0322	-2.2047×10^{-5}	-2.1546×10^{-5}	−0.0292	−0.9998
			Best	−0.2457	-2.9625×10^{-5}	-2.9528×10^{-5}	−0.2493	−1.0000
			Worst	-8.0512×10^{-5}	-1.3124×10^{-5}	-1.0547×10^{-5}	-7.9791×10^{-5}	−0.9987
			Std. Dev	0.0770	3.3992×10^{-6}	3.9198×10^{-6}	0.0746	2.4365×10^{-4}
		30	Mean	−0.0423	-2.1773×10^{-5}	-2.1251×10^{-5}	−0.0265	−0.9998
			Best	−0.2800	-2.8970×10^{-5}	-2.9277×10^{-5}	−0.2467	−1.0000
			Worst	-8.0399×10^{-5}	-1.1655×10^{-5}	-1.0158×10^{-5}	-8.0536×10^{-5}	−0.9983
			Std. Dev	0.0916	3.8563×10^{-6}	3.9806×10^{-6}	0.0723	2.5336×10^{-4}
Friedman mean rank test of the algorithms based on average mean values				3.3	4.1	4.4	2	1.3

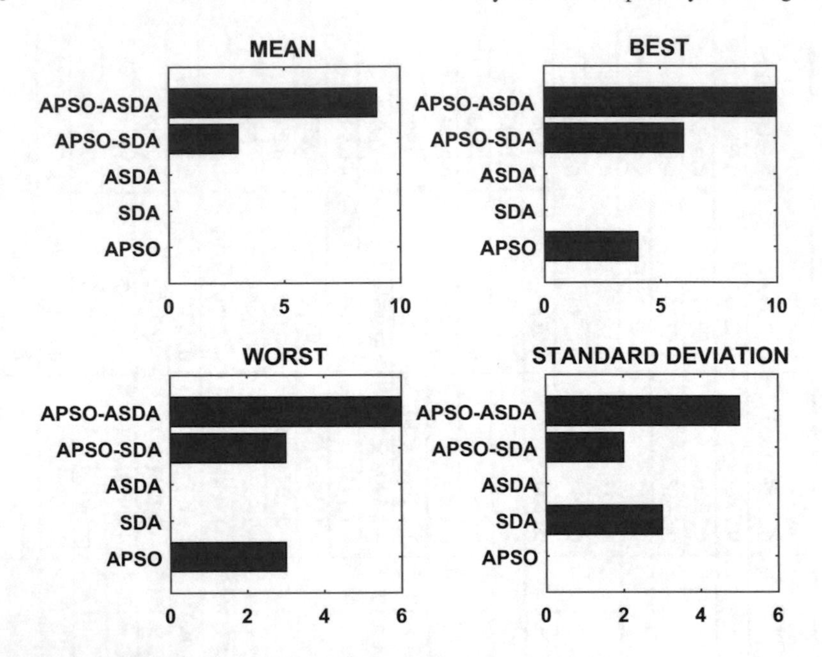

Fig. 4.7 Chart summary for performance of various algorithms

Table 4.5 Summary of Algorithms performance with benchmark functions

	APSO	SDA	ASDA	APSO-SDA	APSO-ASDA
Mean	0	0	0	6	9
Best	4	0	0	6	10
Worst	3	0	0	3	6
Std. Dev.	0	3	0	2	5

pictorial depiction in terms of histogram. It can be seen from both the figure and table that the two proposed algorithms APSO-SDA and APSO-ASDA have the highest frequencies in terms of mean and best. The ranking of algorithms as conducted through Friedman rank test is also provided in Table 4.4. The algorithms best on the test are ranked in the following order: APSO-ASDA, APSO-SDA, APSO, SDA and ASDA. The best rank of 1.3 is achieved by the APSO-ASDA while the least ranked algorithm is ASDA with a value of 4.4 (Tables 4.5 and 4.6).

4.6 Application to Tuning of Filtered PPI (FPPI) Controller

This section presents the application of APSO, SDA, ASDA, APSO-SDA and APSO-ASDA to tuning of FPPI controller presented in Chap. 2 for application in a WirelessHART networked environment. Similar algorithm parameters chosen for the

Table 4.6 Controller parameters

Plant	K_c	T_i	L_p
$G_1(s)$	0.2	2	4
$G_2(s)$	1	1.2	5
$G_3(s)$	1	2	5
$G_4(s)$	1	1.5	10

benchmark functions are used here except N, $birdStep$, m and k_{max} that are chosen as 50, 50, 20 and 50 respectively. These parameters are arrived at after exhaustive test based on trial and error guided by works reported in [8, 9]. In a similar way to the validation with benchmark section, the initial positions of particles and spiral search points are selected randomly.

The controller parameters are tuned using the compared algorithms with the objective of minimising the integral time absolute error (ITAE) given in (4.17). The ITAE is chosen because of its advantages of producing better performance than other indices such as integral absolute error (IAE), integral square error (ISE), mean square error (MSE) etc. This is because the multiplication of the error term with time augments its effect at higher values. This helps in reducing settling time (t_s) and percentage overshoot ($\%OS$) [10].

$$J = \int_{\tau=0}^{t} \tau|e(\tau)|d\tau \tag{4.17}$$

The optimisation structure of the controller is shown in Fig. 4.8. As seen in the figure, the cost ITAE, which is a function of the error, is used in the optimisation algorithm to select controller parameters K_c, T_i and T_f.

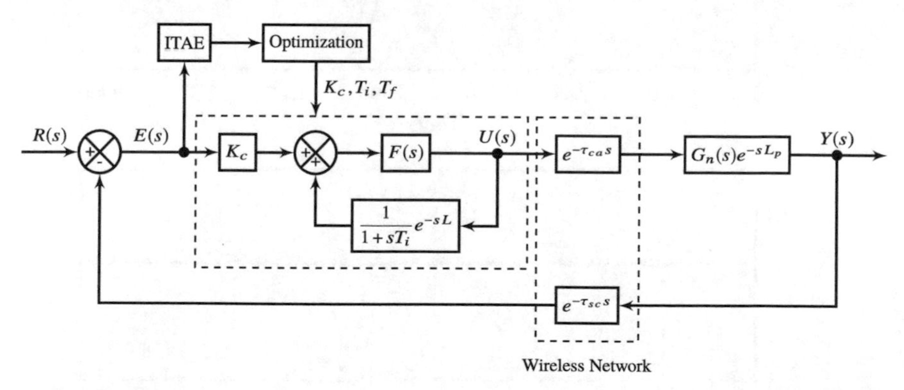

Fig. 4.8 Block diagram of the optimisation structure of the FPPI controller for WirelessHART networked system

4.6.1 Plant Model Selection and Experimental Network Delay

In this work, the plant models given in Eqs. (4.18)–(4.21) are used for simulation with the FPPI controller optimised with APSO, SDA, ASDA, APSO-SDA and APSO-ASDA. These models represent first, second, third and fourth order plus dead time systems as reported in [1, 19].

$$P_1(s) = \frac{5}{1+2s}e^{-4s} \tag{4.18}$$

$$P_2(s) = \frac{1}{(1+1.5s)(1+0.4s)}e^{-5s} \tag{4.19}$$

$$P_3(s) = \frac{1}{s^3+s^2+2s+1}e^{-5s} \tag{4.20}$$

$$P_4(s) = \frac{1}{(1+s)(1+0.5s)(1+0.25s)(1+0.125s)}e^{-10s} \tag{4.21}$$

Figure 4.9 shows the delay between the gateway and a single mote (node) in the WirelessHART network. Mote-to-gateway delay is the upstream delay t_u while the gateway-to-mote delay is the downstream delay t_d. Here, t_u is taken as τ_{sc} while t_d

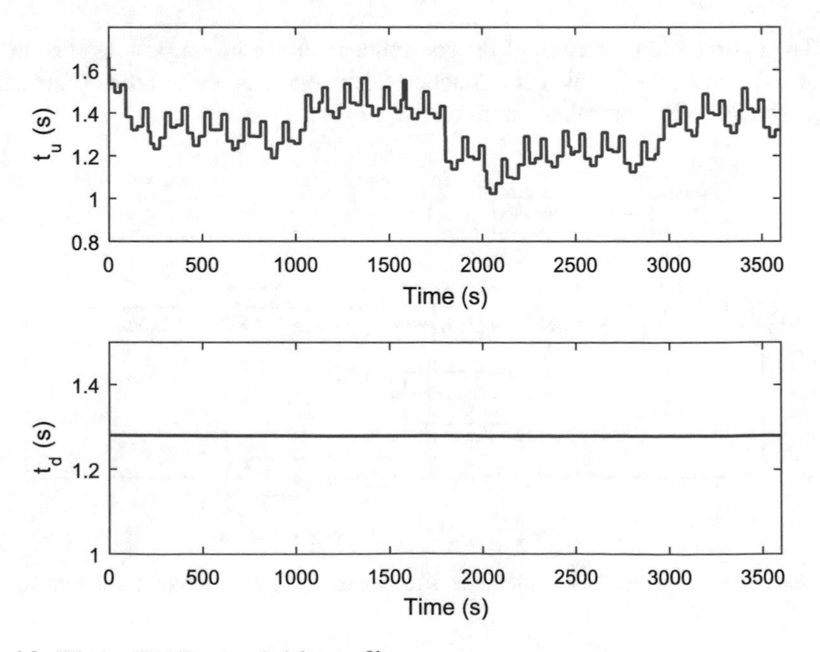

Fig. 4.9 WirelessHART network delay profile

Table 4.7 Statistics of wirelessHART network delay

Delay (s)	Min	Max	Mean	Std.
t_u	1.0220	1.6600	1.3159	0.1173
t_d	1.2800	1.2800	1.2800	0.0000

is taken as τ_{ca} see Fig. 1.8. The stochastic variation of upstream delay over time is as a result of communication between the mote and gateway. To reduce power consumption and preserve battery since usually battery powered, the motes will go into idle state after each communication cycle with the gateway or other motes is completed. In this state, the motes takes longer time to process signal due to lower processing capability. This contributes to the upstream delays variation. On the other hand, the gateway is always in active state since it is the host in the network, hence constant downstream delay.

The average values of the two delays t_d and t_u from the statistical information of Fig. 4.9 given in Table 4.7. The total loop delay L for each plant configuration is calculated based on Eqs. (5.14) and (5.13). Thus the estimate for the L are given as 6.5959, 7.5959, 7.5959 and 12.5959 s for first, second, third and fourth order plants respectively. These values are used to design the FPPI controller for each plant. Each of the four plants is simulated to a step signal of unit magnitude and a disturbance of 50% is injected at the input at 80 s to test for robustness of the controller.

4.6.2 First Order Plant

The tuned controller parameters using the five compared algorithms are given in Table 4.8. The closed loop response of the plant with the controller optimised with these algorithms is shown in Fig. 4.10. The performance analysis with respect to rise time t_r, settling times before and after disturbance t_{s_1}, t_{s_2} and percentage overshoot $\%OS$ is also given in Table 4.8. From the responses as well as the table, the controller with SDA settles faster before disturbance at 13.3204 s. The APSO algorithm produced the fastest response with $t_r = 3.1181$ and settled faster after disturbance at 99.0945 s. However, this is at the cost of very high overshoot of more than 10%. On the other hand, the two hybrid algorithms produced very less overshoots of around 0% each despite slow rise and settling times. Here it is observed that for the FPPI controller, low overshoot values comes at the expense of longer settling and rise times. This can be observed clearly with ASDA and hybrid algorithms. In addition, the controller optimised with the hybrid algorithms give smoother signals compared to the other algorithms.

Table 4.8 Tuned controller parameters and performance analysis for $P_1(s)$

Algorithm	Optimised controller parameters				Plant control performance			
	K_c	T_i	T_f	J_{min}	t_r	t_{s_1}	t_{s_2}	$\%OS$
APSO	0.2241	2.3768	0.1172	716.9085	3.1181	22.7385	99.0945	10.6525
SDA	0.1930	1.9412	0.1576	785.9947	5.2204	13.3204	103.1610	1.6891
ASDA	0.1416	1.3064	0.1945	837.7454	10.0192	21.9902	110.3579	0.4064
APSO-SDA	0.1890	2.2952	0.6218	766.6157	6.3924	21.3873	108.6295	0.0162
APSO-ASDA	0.1842	1.4132	1.2366	781.7953	7.6084	24.5817	111.1369	0.0100

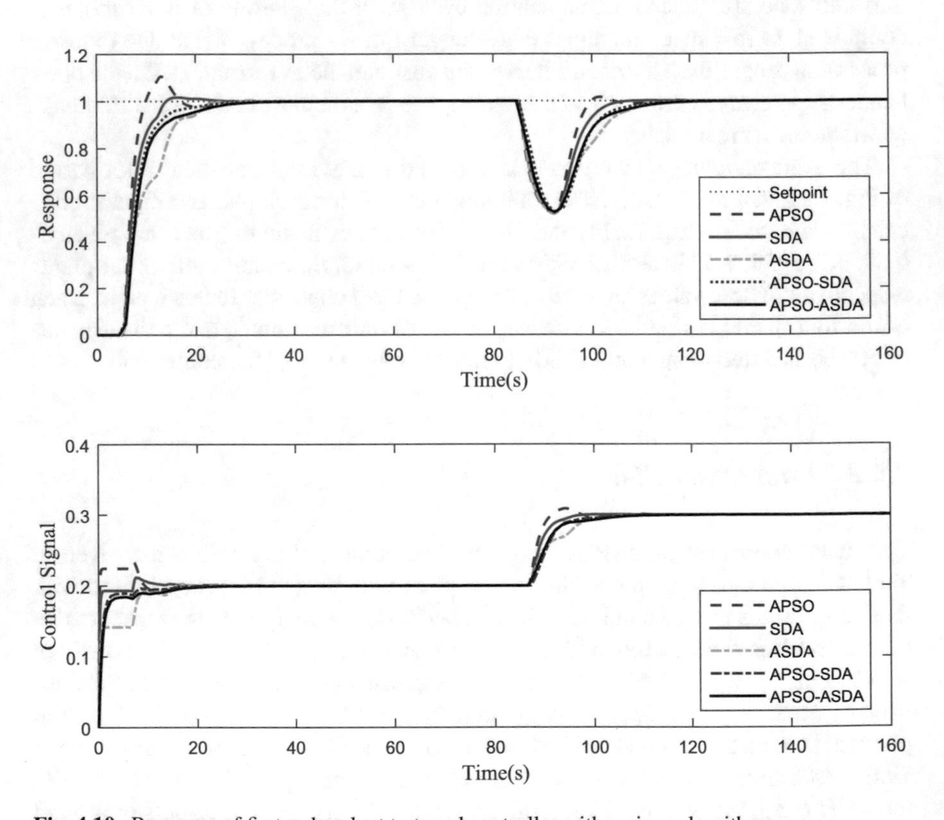

Fig. 4.10 Response of first order plant to tuned controller with various algorithms

Table 4.9 Tuned controller parameters and performance analysis for second order plant

Algorithm	Optimised controller parameters				Plant control performance			
	K_c	T_i	T_f	J_{min}	t_r	t_{s1}	t_{s2}	$\%OS$
APSO	1.0410	1.9019	0.4197	877.4035	3.4562	20.1515	107.4818	10.4740
SDA	1.0858	3.0899	2.1745	1093.5270	5.1132	31.8760	119.8577	5.7149
ASDA	0.7607	1.5853	2.1729	1016.1714	18.1427	41.4342	125.2754	0.0445
APSO-SDA	0.9675	2.1839	1.4097	917.4698	4.8924	17.9435	106.8817	0.0042
APSO-ASDA	0.9821	2.1773	2.4817	1045.6743	7.3575	18.3783	108.1609	0.0024

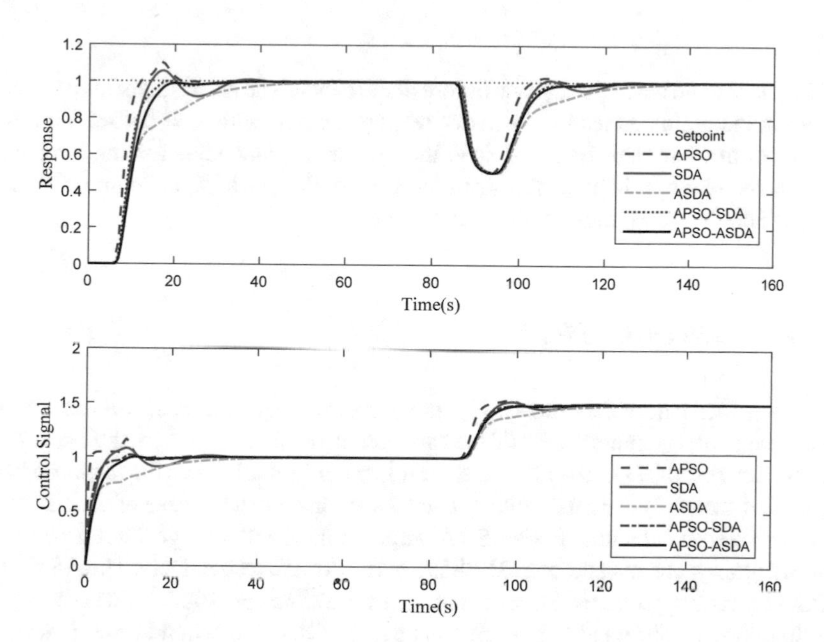

Fig. 4.11 Response of second order plant to tuned controller with various algorithms

4.6.3 Second Order Plant

In a similar fashion to the first order, the controller parameters tuned with the compared algorithms are given in Table 4.9. The closed loop responses of the plant with controller optimised with these algorithms is shown in Fig. 4.11. The numerical analysis of the performance with respect to t_r, t_{s_1}, t_{s_2} and $\%OS$ is also given in Table 4.9. From the figure and the table, it can be seen that unlike in the case of first order, the controller tuned with hybrid algorithms performed better in terms of both settling times and overshoot. For example, best settling times t_{s_1} and t_{s_2} are obtained with the APSO-SDA algorithm, while the overshoot is second best to APSO-ASDA. On the other hand, The APSO-ASDA produced the best overshoot of 0.0024% while

Table 4.10 Tuned controller parameters and performance analysis for third order plant

Algorithm	Optimised controller parameters				Plant control performance			
	K_c	T_i	T_f	J_{min}	t_r	t_{s_1}	t_{s_2}	$\%OS$
APSO	0.9224	2.939	0.4446	1177.5720	6.3791	32.9465	106.7365	5.7461
SDA	1.0558	5.2512	3.1918	1330.5287	7.3374	40.3866	129.1916	1.6319
ASDA	0.9560	4.5164	0.8719	1314.4729	5.6752	34.9922	124.7819	0.0025
APSO-SDA	0.9359	3.7556	1.5513	1310.0233	7.8696	35.0500	109.9817	1.3585
APSO-ASDA	0.9737	3.0986	3.7691	1433.9312	9.8035	21.8819	112.8486	0.0021

achieving second best settling time before disturbance. The APSO algorithm notwithstanding maintains its lead in terms of response speed with $t_r = 3.4562$ s but with overshoot of more than 10% just as in the first order plant. Considering the control signals, the two hybrid algorithms produced smooth control signal compared to the others. This is in congruent with the first order plant.

4.6.4 Third Order Plant

The controller parameters tuned with the compared algorithms as well as the performance analysis numerically of the third order plant are given in Table 4.10. The closed loop responses of the plant is also shown in Fig. 4.12. From the response in the figure and the analysis in the table, it can be seen that as in the case of second order, the controller tuned with APSO-ASDA outperforms the other algorithms in terms of settling time t_{s_1} and overshoot at 21.8819 s and 0.0021% respectively. The ASDA and APSO algorithm produced better t_r and t_{s_2} of 5.6752 s and 106.7365 s respectively. Furthermore, as observed in the previous cases of first and second order plants, the action of the controller optimised with APSO-ASDA is smoother compared to the other algorithms. This consequently makes the tracking of the controller better as observed from the response in the figure and the table.

4.6.5 Fourth Order Plant

The controller parameters of the fourth order plant tuned with the compared algorithms as well as the performance analysis numerically are given in Table 4.11. The closed loop responses of the plant is also shown in Fig. 4.13. From the response in the figure and the analysis given in the table, it can be seen that the controller tuned with APSO-ASDA better settling times t_{s_1} and t_{s_1} at 27.3603 s and 120.7414 s respectively. This controller also gives moderate overshoot of less than 1%. Although

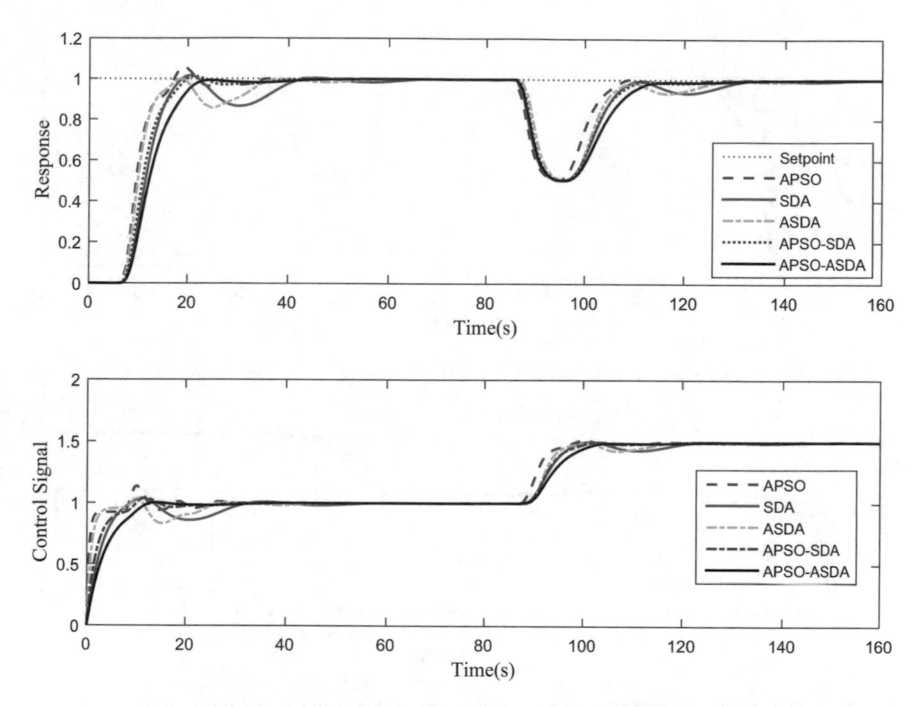

Fig. 4.12 Response of third order plant to tuned controller with various algorithms

Table 4.11 Tuned controller parameters and performance analysis for fourth order plant

Algorithm	Optimised controller parameters				Plant control performance			
	K_c	T_i	T_f	J_{min}	t_r	t_{s_1}	t_{s_2}	$\%OS$
APSO	0.9129	2.03175	0.7178	1660.0668	5.3735	36.3280	128.2603	1.9527
SDA	0.9307	2.6769	1.1543	1790.1409	5.3851	39.5128	131.9215	0.2029
ASDA	0.9114	2.4488	2.6162	1896.3266	11.6624	43.6503	134.2368	0.1386
APSO-SDA	1.0368	2.2074	5.0886	2032.7493	10.2803	27.3998	121.1694	1.1706
APSO-ASDA	0.9737	2.3240	3.7691	1983.2289	9.8302	27.3603	120.7414	0.8374

the ASDA and SDA produced better overshoots of 0.1386% and 0.2029% respectively, their responses is however not as smooth as the response of APSO-SDA and APSO-ASDA. Furthermore, the APSO and SDA produced faster response with t_r of approximately 5.3 s each. Despite slightly higher overshoots of the APSO-SDA and APSO-ASDA optimised controllers, their control action is smoother compared to the others.

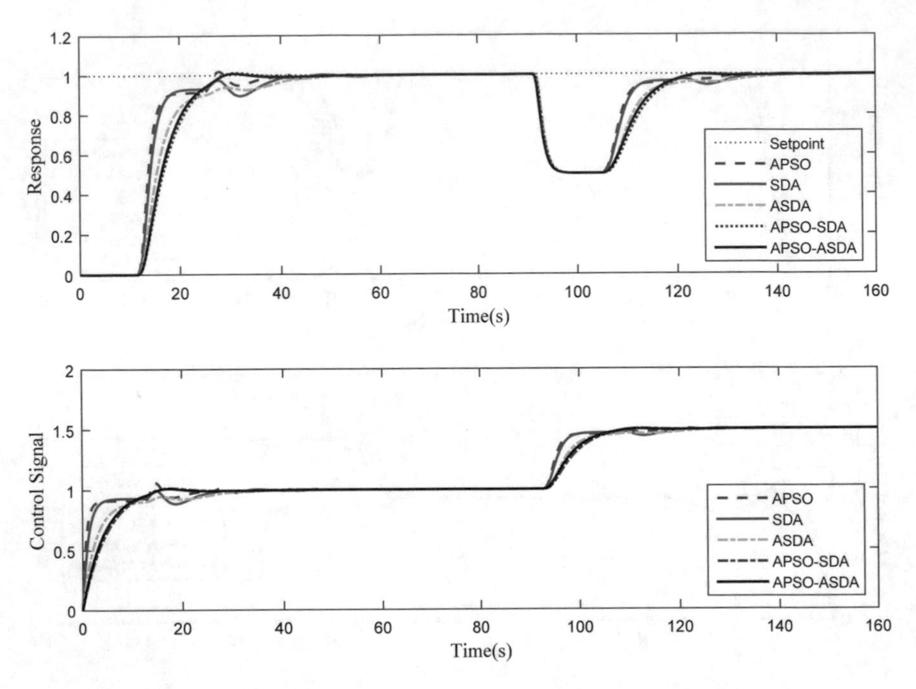

Fig. 4.13 Response of fourth order plant to tuned controller with various algorithms

4.7 Summary

In this chapter, two hybrid algorithms synergizing the APSO with SDA on one hand and APSO with ASDA on the other hand have been proposed. In the proposed algorithms, the best solutions of the APSO are used as initial search points of either the SDA or the ASDA to achieve APSO-SDA and APSO-ASDA respectively. Performance of the proposed algorithms in comparison to their constituents algorithms has been evaluated using several benchmark functions. Mean rank test conducted on the compared algorithms through Friedman's rank test suggests that the algorithm with the best rank is APSO-ASDA. This is followed by APSO-SDA, then APSO. The last two in the ranking are the SDA and ASDA. This ranking result indicates that the proposed algorithms have improved on their constituent algorithms. The proposed hybrid algorithms have been used to tune parameters of FPPI controller designed for application in a wireless networked environment. Results have shown that the proposed algorithms produce controller with good time domain performance of faster settling times as well as less overshoot compared to constituent algorithms.

References

1. Nasir, A.N.K., Tokhi, M.O.: A novel hybrid bacteria-chemotaxis spiral-dynamic algorithm with application to modelling of flexible systems. Eng. Appl. Artif. Intell. **33**, 31–46 (2014). Elsevier
2. Esmin, A.A., Coelho, R.A., Matwin, S.: A review on particle swarm optimization algorithm and its variants to clustering high-dimensional data. Artif. Intell. Rev. **44**(1), 23–45 (2015). Springer
3. Yang, X.S.: Nature-Inspired Optimization Algorithms. Elsevier, Amsterdam (2014)
4. Yang, X.S.: Engineering Optimization: An Introduction with Metaheuristic Applications. Wiley, New York (2010)
5. Passino, K.M.: Biomimicry of bacterial foraging for distributed optimization and control. IEEE Control Syst. **22**(3), 52–67 (2002). IEEE
6. Dorigo, M., Birattari, M., Stutzle, T.: Ant colony optimization. IEEE Comput. Intell. Mag. **1**(4), 28–39 (2006). IEEE
7. Tamura, K., Yasuda, K.: Primary study of spiral dynamics inspired optimization. IEE J. Trans. Elect. Electron. Eng. **6**(1), 98–100 (2011). Wiley Online Library
8. Tamura, K., Yasuda, K.: Spiral dynamics inspired optimization. J. Adv. Comput. Intell. Intell. Inf. **15**(8), 1116–1122 (2011)
9. Yang, X.S, Deb, S., Fong, S.: Accelerated particle swarm optimization and support vector machine for business optimization and applications. Int. Conf. Netw. Digital Technol. 53–66 (2011). Springer
10. Sahib, M.A., Ahmed, B.S.: A new multi-objective performance criterion used in PID tuning optimization algorithms. J. Adv. Res. **7**(1), 125–134 (2016). Elsevier
11. Mirjalili, S., Lewis, A.: The whale optimization algorithm. Adv. Eng. Softw. **95**, 51–67 (2016). Elsevier
12. Zhou, Y., Ling, Y., Luo, Q.: Lévy flight trajectory-based whale optimization algorithm for global optimization. IEEE Access. (2017). IEEE
13. Muthukumar, V., Babu, A.S., Venkatasamy, R., Kumar, N.S.: An accelerated particle swarm optimization algorithm on parametric optimization of WEDM of die-steel. **96**(1), 49–56 (2015). Springer
14. Subha, R., Himavathi, S.: Accelerated particle swarm optimization algorithm for maximum power point tracking in partially shaded PV systems. In: 3rd International Conference on Electrical Energy Systems (ICEES), pp. 232–236 (2016). IEEE
15. Paschos, A.E., Kapinas, V.M., Hadjileontiadis, L.J., Karagiannidis, G.K.: Dynamic spectrum sensing using a novel accelerated particle swarm optimization algorithm (2015). arXiv:1510.03840
16. Wang, G., Hossein, G.A., Yang, X.S., Hossein, A.A.: A novel improved accelerated particle swarm optimization algorithm for global numerical optimization. Eng. Comput. **31**(7), 1198–1220 (2014). Emerald Group Publishing Limited
17. Guedria, N.B.: Improved accelerated PSO algorithm for mechanical engineering optimization problems. Appl. Soft Comput. **40**, 455–467 (2016). Elsevier
18. Nasir, A.N.K., Tokhi, M.O., Ghani, N.M.A., Ismail, R.M.T.: Novel adaptive spiral dynamics algorithms for global optimization. In: Proceeding of the 11th IEEE International Conference on Cybernetics Intelligent System, pp. 99–104 (2012). IEEE
19. Nasir, A.N.K., Tokhi, M.O., Sayidmarie, O., Ismail, R.M.T.: A novel adaptive spiral dynamic algorithm for global optimization. In: 13th UK Workshop on Computational Intelligence (UKCI), pp. 334–341 (2013). IEEE
20. Nasir, A.N.K., Tokhi, M.O., Omar, M.E., Ghani, N.M.A.: An improved spiral dynamic algorithm and its application to fuzzy modelling of a twin rotor system. In: 2014 World Symposium on Computer Applications & Research (WSCAR), pp. 1–6 (2014). IEEE
21. Nasir, A.N.K., Tokhi, M.O.: An improved spiral dynamic optimization algorithm with engineering application. IEEE Trans. Syst. Man Cybern. Syst. **45**(6), 943–954 (2015)

22. Nasir, A.N.K., Tokhi, M.O.: Novel metaheuristic hybrid spiral-dynamic bacteria-chemotaxis algorithms for global optimisation. Appl. Soft Comput. **27**, 357–375 (2015). Elsevier
23. Jamil, M., Yang, X.S: A literature survey of benchmark functions for global optimisation problems. Int. J. Math. Model. Numer. Optim. **4**(2), 150–194 (2013). Inderscience Publishers Ltd
24. Butler, A., Haynes, R.D., Humphries, T.D., Ranjan, P.: Efficient optimization of the likelihood function in Gaussian process modelling. Comput. Stat. Data Anal. **73**, 40–52 (2014). Elsevier

Chapter 5
Hybrid ABFA-APSO Algorithm

5.1 Introduction

The aim of this chapter is to propose improvement to the adaptation of bacterial foraging algorithm (BFA) and to hybridize it with accelerated particle swarm optimization (APSO) in order to accelerate its convergence. In the proposed algorithm, the random walk in the chemotaxis stage of the ABFA is updated through the velocity equation of the APSO. This in turn accelerates the convergence of the algorithm. The algorithm will be validated using selected benchmark functions. Subsequently, an optimal fuzzy PID controller for application in a WirelessHART networked control environment characterized by stochastic network delay will be designed. The controller parameters will be selected using the proposed algorithms.

The next section of the paper will present related work on bacterial foraging optimisation. This will be followed by a brief discussion on the standard BFA algorithm as well as its adaptive variants. The next section of the chapter will hybridize the BFA with APSO algorithm. Consequently, the fuzzy PID controller will be designed and tuned for a class of first and second order systems using the proposed algorithms. Subsequently, simulation results comparing the controller tuned with proposed algorithms as well as constituent algorithms will be provided. Lastly, a summary will be given at the end of the chapter.

5.2 Related Works on Bacterial Foraging Optimisation

The BFA is developed based on the mimicry of the foraging strategy of *Escherichia coli* (E. *coli*) bacteria which is found commonly in the intestine of endotherms such as birds and mammals including humans [1]. Here, the optimisation strategy adopted by the E. *coli* is adopted as the basis for the algorithm. This algorithm was chosen in this work because compared to the two widely used algorithms GA and PSO, it has overcome the problem of local minima which leads to premature convergence in

S. M. Hassan et al., *Hybrid PID Based Predictive Control Strategies for WirelessHART Networked Control Systems*, Studies in Systems, Decision and Control 293,
https://doi.org/10.1007/978-3-030-47737-0_5

both GA and PSO [1–7]. This is because of its unique features of Elimination and Dispersal event which enables the algorithm to find good solution even with small population of the bacteria. This phenomenon has the possibility of placing bacteria where there is high nutrient value. Furthermore, the algorithm requires a moderate memory for its implementation [1].

A key parameter in the algorithm that determines how the bacteria moves in any nutrient profile is the chemotactic step length. If the length is too short, the algorithm takes longer time to converge. On the other hand, if it is too long there is the possibility that the bacteria may skip a region of high nutrient (i.e. optimal position), hence will end up getting stuck in local minima. Moreover, the standard algorithm proposes a fixed chemotactic step length which is impractical in a dynamically changing nutrient profile environment. To overcome this problem, several adaptation strategies have been proposed by researchers. The use of strategies such as the principle of adaptive delta modulation [8] and special adaptation in [9, 10] schemes were proposed. However, these two strategies only add to the computation complexity of the algorithm. This is because, in each of the two cases, additional algorithms have to be computed before determining the new step size that will be used in the computation of the BFA.

The use of linear, quadratic and exponential functions to dynamically adapt the chemotactic step length have also been proposed for modeling of dynamic system and other engineering applications [2–4, 11–13]. In all these, the chemotactic step length is adjusted based on the linear, quadratic or exponential function of the nutrient profile. This implies that a change in the nutrient profile produces a corresponding change in the step size. A key problem with these functions is that whenever the nutrient profile goes to zero, the step size also goes to zero. This scenario will constitute a serious problem for the algorithm. Since according to the original BFA, the step size cannot be zero. Also, the chemotactic step is taken randomly.

Some attempts were made to hybridize the algorithm by combining the features of the BFA and other popular algorithms such as GA [5, 14], Differential Evolution (DE) [6], PSO [15] and spiral dynamic algorithm (SDA) [16, 17]. All these attempts are to improve the performance of the standard algorithm by making it adaptive to dynamic environments and to ensure faster convergence. However, no attempt has been made to take advantage of the fast convergence of the APSO algorithm by hybridizing it with the variants of the BFA algorithms.

5.3 Bacterial Foraging Optimisation

The chemotactic (foraging) strategy of E. *coli* bacteria conforms to the Natural selection principle of favoring the propagation of genes of species with successful foraging strategy [1]. The already evolved foraging strategy of E. *coli* is characterized by four main actions i.e. chemotaxis, swarming, reproduction and elimination-dispersal [1, 16, 18] as explained subsequently.

5.3.1 Chemotaxis

This process is characterized by swimming and tumbling. This can been explained as the movement of a bacterium by rotating its flagella in a predefined way (swimming) or in altogether different ways (tumbling). In the algorithm, a tumble is represented using a unit length $\phi(j)$ which in turn is used to define the movement direction of the bacterial. The position θ after $(j+1)$th chemotactic movement of the ith bacteria is thus defined as

$$\theta^i(j+1,k,l) = \theta^i(j,k,l) + C(i)\phi(j) \tag{5.1}$$

where k and l represent the steps of reproduction and elimination/dispersal respectively. $C(i)$ is the chemotactic step length. Thus, the bacteria advances in the same direction by one step if the cost $J(i, j+1, k, l)$ at $\theta^i(j+1, k, l)$ is less than $J(i, j, k, l)$ at (θ^i, k, l). This continues for a maximum step N_s.

5.3.2 Swarming

The ability of a bacterium in the population to attract a neighbor bacterium by releasing an attractant for the purpose of swarming together towards the optimal position of nutrient is called swarming. This swarming effect is similar to social behavior in other animals that exhibit group foraging. The combine swarm effect between the bacteria cell can be denoted as

$$J_{cc} = \sum_{i=1}^{S} J_{cc}^i(\theta, \theta^i(j,k,l)) \tag{5.2}$$

where $\theta = [\theta_1, \theta_2, \ldots, \theta_p]$ is a given point on the search space. Now, the combine cost of the ith bacterium for swarming is given as

$$J_{total} = J(i, j, k, l) + J_{cc} \tag{5.3}$$

5.3.3 Reproduction

In the process of reproduction, the healthy bacterium each split into two new bacteria, while the least healthy bacteria die. By this action the population of the bacteria is maintained throughout the foraging process. Hence, after chemotactic steps N_c are taken, a reproduction step is taken. This process continues until a maximum reproduction step N_{re} is reached. Thus, if S is the total bacteria, the number of healthy bacteria is given as

$$S_r = \frac{S}{2} \tag{5.4}$$

such that after each reproduction the population S is maintained.

5.3.4 Elimination and Dispersal

As the bacteria forages, there is tendency that some bacteria maybe placed in a noxious nutrient profile or that some bacteria maybe placed into new environment with high nutrients. This phenomenon is referred as elimination-dispersal event. This event may occur either as a result of consumption of the nutrients or via other means such as heat, wind or even animals [1].

5.3.5 Bacterial Foraging Algorithm Flowchart

The flowchart which implements the foraging strategy of the E. *coli* bacteria is given in Fig. 5.1. The flowchart is made up loops implementing the chemotactic, reproduction, and elimination and dispersal events for each bacterium in the population. Similar flowchart is used for implementation of the adaptive BFA (ABFA).

5.3.5.1 Standard Bacterial Foraging Algorithm (BFA)

In the standard BFA, the key parameter that controls the search ability of each bacterium is the fixed chemotactic step length $c(i)$. If this fixed length is too short, the algorithm takes longer time to converge. On the other hand, if it is too long there is the possibility that the bacteria may skip a region of high nutrient (i.e. optimal position), hence will end up getting stuck in a local minima. In this work, an adaptive chemotactic step length is proposed for both the standard BFA and the hybrid BFA-PSO algorithms. The proposed adaptive step length to be used in the ABFA and both HABF-PSO and HABF-APSO is a function of the nutrient profile presented in (5.7). If the cost is increasing (low nutrient), the step size approaches c_{max}, and the bacteria takes a larger step to find nutrient. On the other hand, if the cost is decreasing (High nutrient), the bacteria takes a shorter step to consume the nutrient and to settle at the global minimum. This is expected to make the algorithm converge faster

$$c_a(i) = \frac{c_{max}}{1 + \dfrac{\alpha}{\left|\beta + J^i(\theta)\right|}} \tag{5.5}$$

where, α and β are tunable positive parameters, c_{max} is chosen based on the nature of the nutrient profile function and $J^i(\theta)$ is the cost for the ith bacterium.

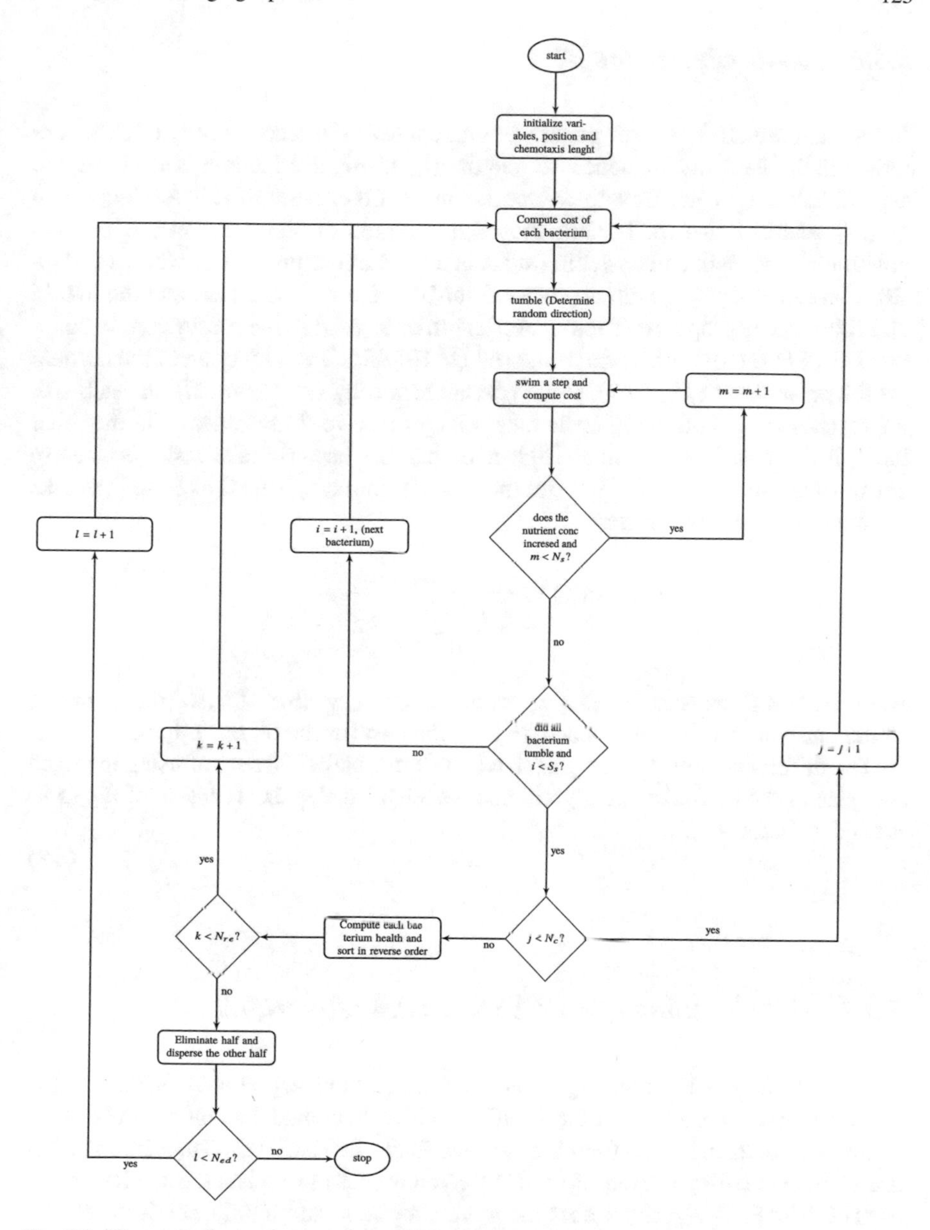

Fig. 5.1 Flowchart for implementation of BFA and ABFA

The difference between the adaptive algorithms is that instead of using the fixed step size of the non adaptive algorithms, the adaptive stepsize function of Eq. (5.8) is used as follows:

$$c(i) = c_a(i). \tag{5.6}$$

5.3.6 Adaptation to the BFA

In the standard BFA, the key parameter that controls the search ability of each bacterium is the fixed chemotactic step length $c(i)$. If this fixed length is too short, the algorithm takes longer time to converge. On the other hand, if it is too long there is the possibility that the bacteria may skip a region of high nutrient (i.e. optimal position), hence will end up getting stuck in a local minimum. In this work, an adaptive chemotactic step length is proposed for both the standard BFA and the hybrid algorithms. The proposed adaptive step length to be used in the ABFA, hybrid adaptive BF-PSO (HABF-PSO) and proposed HABF-APSO is a function of the nutrient profile presented in Eq. (5.7). If the cost is increasing (low nutrient), the step size approaches c_{max}, and the bacteria take a larger step to find nutrient. On the other hand, if the cost is decreasing (High nutrient), the bacteria take a shorter step to consume the nutrient and to settle at the global minimum. This is expected to make the algorithm converge faster

$$c_a(i) = \frac{c_{max}}{1 + \dfrac{\alpha}{\left|\beta + J^i(\theta)\right|}} \tag{5.7}$$

where, α and β are tunable positive parameters, c_{max} is chosen based on the nature of the nutrient profile function and $J^i(\theta)$ is the cost for the ith bacterium.

The difference between the adaptive algorithms is that instead of using the fixed step size of the non-adaptive algorithms, the adaptive stepsize function of Eq. (5.8) is used as follows:

$$c(i) = c_a(i). \tag{5.8}$$

5.3.7 Hybridization of ABFA with APSO Algorithm

The proposed HABF-APSO algorithm, just like its counterpart HABF-PSO, makes use of the social behaviour of the APSO algorithm to update the random walk of the chemotaxis of the bacterial foraging process. While the HABF-PSO algorithm makes use of the velocity equation of the PSO given in (5.9) to update the random walk, the HABF-APSO algorithm uses the more simplified and highly efficient velocity equation of the APSO algorithm also given in (5.10). This makes the algorithm converges faster. The flowchart of the two algorithms is given in Fig. 5.2.

$$V_i(t+1) = \omega V_i(t) + c_1\varphi_1(X_i(t) - x_i(t)) + c_2\varphi_2(X_g(t) - x_i(t)) \tag{5.9}$$

$$V_i(t+1) = V_i(t) + \rho\epsilon_n + \sigma(X_g(t) - x_i(t)) \tag{5.10}$$

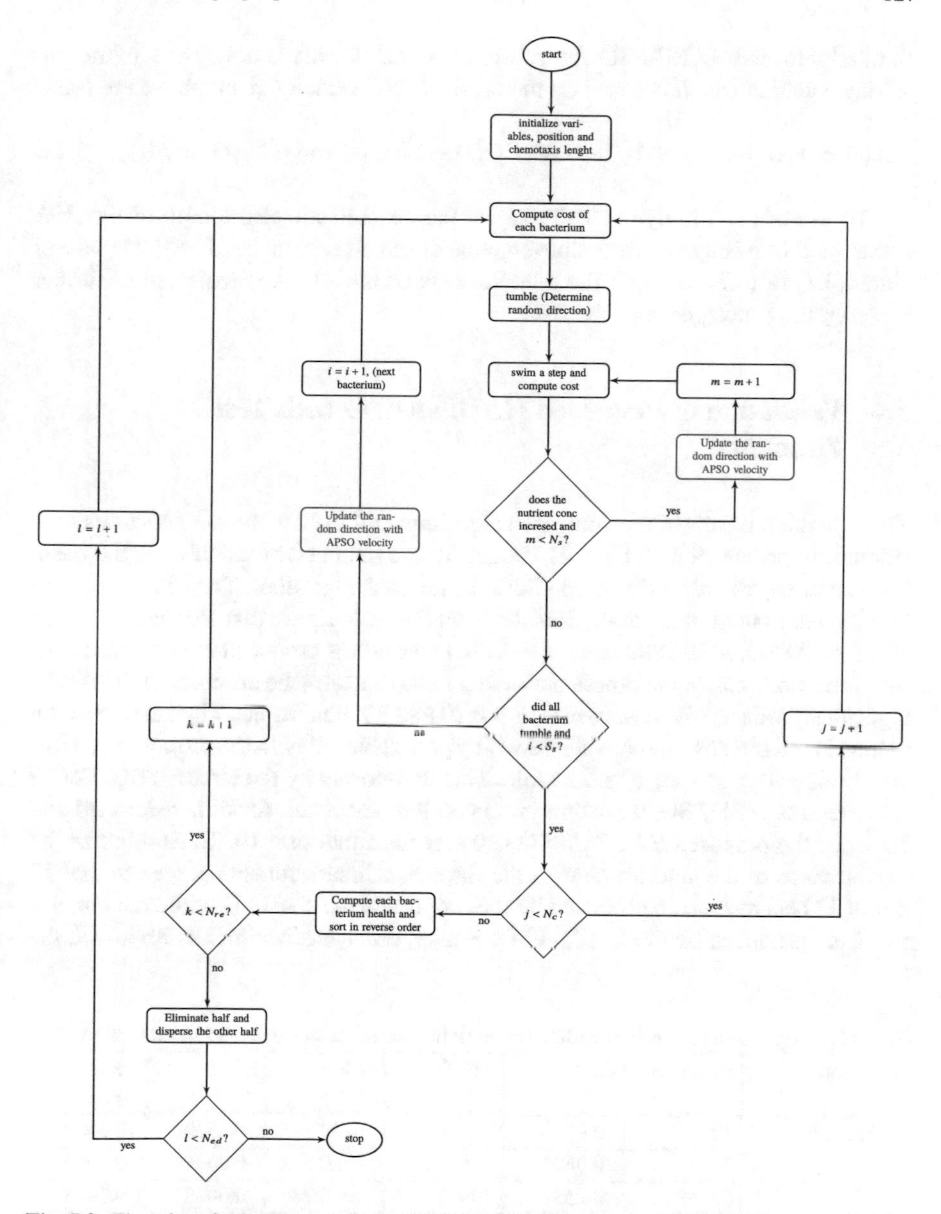

Fig. 5.2 Flowchart for implementation of HABF-APSO algorithm

Thus, for the HABF-APSO algorithm, the random walk is now directed by the velocity equation as follows

$$\Delta(i) = V_i(t+1) = V_i(t) + \rho\epsilon_n + \sigma(X_g(t) - x_i(t)) \tag{5.11}$$

Similarly, for the HABF-PSO algorithm, the directed walk is now given in the following equation which is based on the original PSO velocity given above see (4.1)

$$\Delta(i) = V_i(t+1) = \omega V_i(t) + c_1\varphi_1(X_i(t) - x_i(t)) + c_2\varphi_2(X_g(t) - x_i(t)) \quad (5.12)$$

The use of (5.7) to dynamically adjust the chemotactic step-length of the BFA is expected to result in a faster convergence of the algorithm. Similarly, The use of either (4.1) or (4.3) to direct the random walk of the ABFA is expected to further speed up the convergence of the BFA.

5.4 Validation of Proposed HABF-APSO with Test Functions

The validation results of the compared algorithms for the benchmark functions considered are presented in Table 5.1. The results are obtained by calculating the mean of the optimal values of 50 simulation runs for each algorithm. The best results are highlighted in bold in the table. Results from the table shows that for the dimension 10 of f_1, APSO, ABF-PSO and ABF-APSO algorithms produced the best results of the global optimum of the function. For this same function, the proposed ABF-APSO algorithm produced the best mean for the 20 and 30 dimensions. The best optimum means for all dimensions considered with f_2 are achieved with the proposed HABF-APSO algorithm as shown in the table. This is followed by the HABF-PSO. For f_3 however, the ABF-PSO algorithm produced the best result for dimensions 20 and 30 while the proposed HABF-APSO is better for dimension 10. To summarize the performance of the algorithms with the three benchmark functions, we see that 10 out of 12 best results are produced by the proposed HABF-APSO while the remaining 2 are produced by the HABF-PSO. Hence, the hybridization has improved the

Table 5.1 Algorithm verification with benchmark functions: mean values (50 simulation runs)

Function	Dimension	PSO	APSO	ABFA	HABF-PSO	HABF-APSO
Sphere (f_1)	10	0.0045	0.0000	0.0059	0.0000	0.0000
	20	0.0478	0.0921	0.0215	0.0004	0.0003
	30	0.2538	0.4121	0.0407	0.0009	0.0005
Ackley (f_2)	10	0.0681	0.0529	0.1386	0.0080	0.0035
	20	0.2096	0.6760	0.2126	0.0133	0.0045
	30	0.5869	1.1559	0.2395	0.0154	0.0036
Rosenbrook (f_3)	10	8.4058	7.5220	9.6341	6.8003	6.4753
	20	25.5478	28.6269	25.3715	16.7799	20.5462
	30	89.6182	70.2625	40.8100	27.2178	47.6061

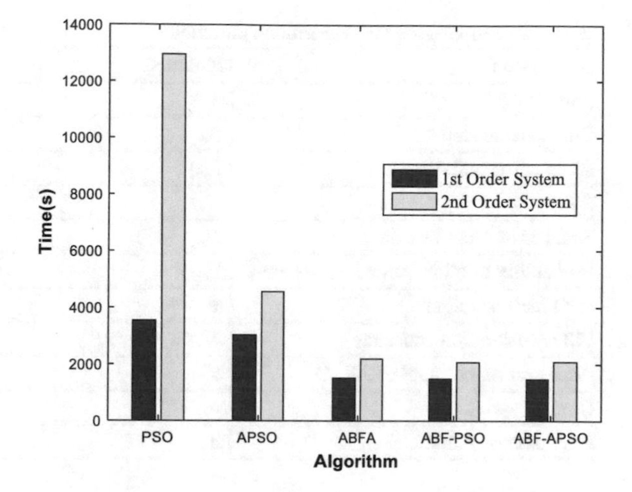

Fig. 5.3 Comparison for simulation times of various algorithms

performance of the individual algorithms. The computation time for this algorithm for first and second order systems are also given in (Fig. 5.3).

5.4.1 *Performance Measure*

The objective cost function adopted for the tuning of the controller parameters is the integral time absolute error (ITAE). This is because the ITAE produces better performance than other indices with higher powers of error and time by reducing the settling time (t_s) and the percentage overshoot (M_p). The function is however modified in (5.13) with a square of the input term to minimise the need for a large control signal to avoid actuator saturation.

$$J = w_1 \int t\,|e(t)|\,\mathrm{d}t + w_2 u^2(t) \tag{5.13}$$

where, w_1 and w_2 are the weighting factors, $u(t)$ is the control signal and $e(t)$ is the error. The bigger the choice of either of the weights w_1 and w_2, the better the suppression of cost term associated with it.

5.4.2 *Algorithms Parameter Selection*

Selection of algorithm parameters in this work is guided by works reported in [16, 19, 20]. The parameters are selected to ensure fairness among the compared algorithms. Thus, the number of fitness evaluation (FE) is defined to be approximately 10,000 for the PSO/APSO and 10240 for the ABFA based algorithms respectively. For the PSO and APSO algorithms, the FE is given as $S \times Iter = 100 \times 100 = 10{,}000$.

Table 5.2 Proposed hybrid algorithms parameters

Description	Parameter	Value
Population	S	32
Chemotactic step	N_c	20
Elimination and dispersal	N_{ed}	4
Swarm	N_s	4
Number of reproduction	N_{re}	4
Probability of elimination and dispersal	P_{ed}	0.25
PSO inertia weight	ω	0.9
PSO acceleration constants	c_1, c_2	1.49
Acceleration constant of APSO	σ	0.7
Control variable	γ	0.7
Randomness parameter	ρ	0.5

On the other hand, the ABFA based algorithms FE is defined by the product $S \times Nc \times Nre \times Ned = 32 \times 20 \times 4 \times 4 = 10240$. Other parameters such as bacterial position x_i, φ_1, φ_2 and ϵ are selected randomly. The adaptive steplength $c_a(i)$ parameters α and β are both chosen as unity while c_{max} is chosen as 0.05.

The advantage of the proposed algorithm is that its based on the dynamism of the bacterial foraging. Thus, good solutions can be obtained even with a small population of the bacteria. Also, the use of the velocity of the APSO accelerates the convergence of the algorithm. The table of the main parameters of the proposed algorithms is given in Table 5.2. These parameters are used to tune both fuzzy PID and the FPPI.

5.5 Implementation of Optimal Fuzzy PID Controller for WHNCS

This section gives the fuzzy logic PID controller design for WirelessHART environment. First, the fuzzy control structure is discussed. Next, discussion on the selection of membership functions follows. Finally, the rule base and the rule base table is presented.

5.5.1 Controller Structure and Tuning

The Fuzzy PID control structure shown in Fig. 5.4 is used in this work. It consists of a PID like Fuzzy controller, the WirelessHART networked variable delays, and the plant. As seen from the structure, the controller has two inputs (error and change in error) and one output. The input scaling factors (SFs) K_e and K_{ce} are the respective

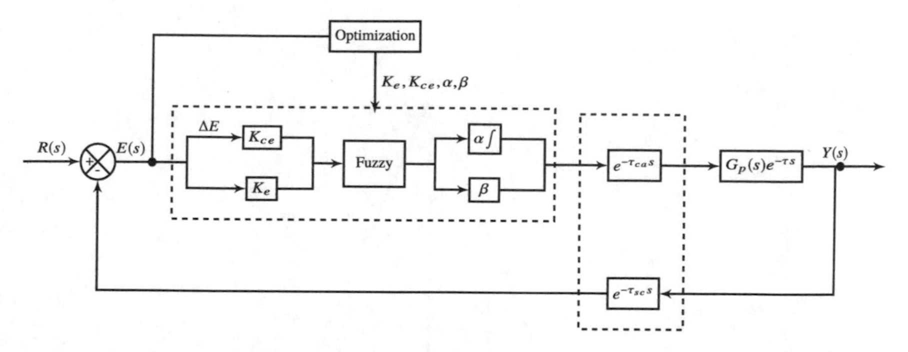

Fig. 5.4 Optimal fuzzy PID structure

error and error change gains, while the output SFs α and β are the control gains. The fuzzy PI is adjudged to be more practical than the Fuzzy PD because of the difficulty to eliminate steady-state error with the Fuzzy PD. However, the fuzzy PI gives poor transient response for higher order systems as a result of internal integration. Thus the Fuzzy PID is more useful in practice. The output of the Fuzzy PID controller shown in Fig. 5.4 is given in Eq. (5.14):

$$u(t) = K_p + K_I \int e(t)\mathrm{d}t + K_d \frac{\mathrm{d}e(t)}{\mathrm{d}t} \tag{5.14}$$

In (5.14), K_p, K_I, and K_d are the proportional, integral, and derivative constants respectively. They are related to the fuzzy PID gains as follows:

$$K_p = \alpha K_{ce} + \beta K_e \tag{5.15}$$

$$K_I = \alpha K_e \tag{5.16}$$

$$K_d = \beta K_{ce} \tag{5.17}$$

where,

- K_e is the input error scaling factor
- K_{ce} is the input error change scaling factor
- α and β are the output scaling factors.

As can be seen, the objective cost function adopted for the tuning of the controller parameters is the same integral time absolute error (ITAE) of (5.13) as explained earlier. The justification for the choice of this function has been given in that section.

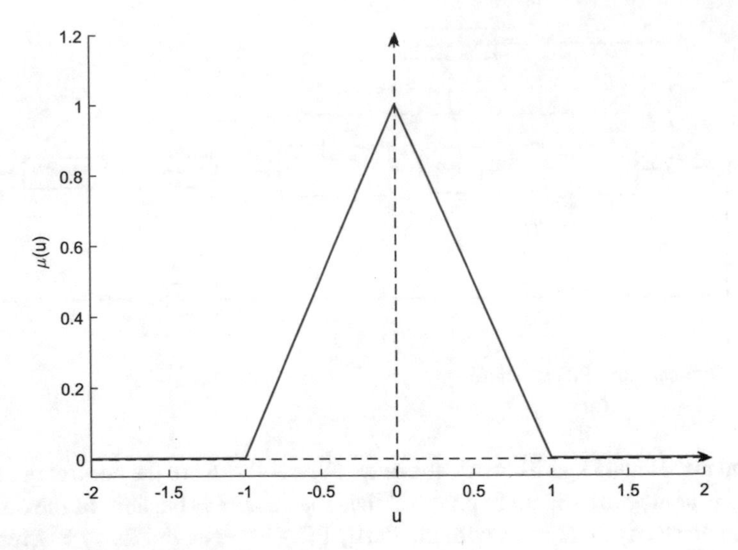

Fig. 5.5 Triangular membership function

5.5.2 Membership Functions

Mamdani type inference system is adopted for the controller design. The method was chosen due to its wide acceptance, intuitiveness and its suitability to human input [19]. A two-dimensional linear rule base with triangular membership function (MF) represented by (5.18) is used for the error (E), error change (ΔE) and the output (UFLC). The pictorial representation of the MF is given in Fig. 5.5. The centre-of-gravity CoG defuzzification method is used to determine the output UFLC.

$$\mu(u; a, b, c) = \begin{cases} 0, & u \le a \\ \frac{u-a}{b-a}, & a \le u \le b \\ \frac{c-u}{c-b}, & b \le u \le c \\ 0, & c \le u \end{cases} \tag{5.18}$$

where a, b, and c are -1, 0 and 1 respectively.

5.5.3 Rule Base

For the fuzzy system, seven linguistic variables namely: Negative Big (NB), Negative Medium (NM), Negative Small (NS), Zero (Z) Positive Small (PS), Positive Medium (PM) and Positive Big (PB) are used for the inputs and output to form the forty-

Table 5.3 Fuzzy PID rule base

E	Δ E						
	NB	NM	NS	Z	PS	PM	PB
PB	Z	PS	PM	PB	PB	PB	PB
PM	NS	Z	PS	PM	PB	PB	PB
PS	NM	NS	Z	PS	PM	PB	PB
Z	NB	NM	NS	Z	PS	PM	PB
NS	NB	NB	NM	NS	Z	PS	PM
NM	NB	NB	NB	NM	NS	Z	PS
NB	NB	NB	NB	NB	NM	NS	Z

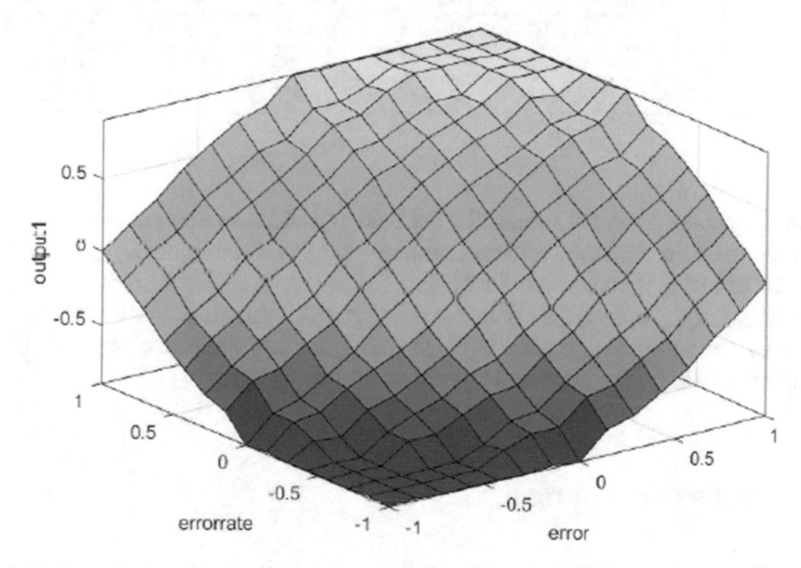

Fig. 5.6 Output surface map

nine (49) rule base as shown in Table 5.3. The output surface map of the controller corresponding to these rules is shown in Fig. 5.6.

5.6 Simulation Results for Optimal Fuzzy PID Control

For this section, two plants namely first order with deadtime (FODT) and second order with deadtime (SODT) given Eqs. (2.34) and (2.35) are considered for simulation. The closed-loop control performance with fuzzy PID controllers tuned with PSO, APSO, ABFA, ABFA-PSO, and ABF-APSO are compared. The initialisation parameters of the algorithms are discussed in Sect. 5.4.2. Table 5.4 shows tuned

Table 5.4 Controller parameters

System	Algorithm	Controller gains				J_{min}
		K_e	K_{ce}	α	β	
First order	HABF-APSO	0.4423	0.1953	0.1913	0.7935	$1.3146e^4$
	HABF-PSO	0.2512	0.3719	0.4807	0.4417	$1.4706e^5$
	PSO	0.6964	0.5635	0.1879	0.1879	$1.1196e^5$
	APSO	0.8338	0.5918	0.1718	0.5810	$1.1543e^5$
	ABFA	0.2247	0.5243	0.3417	0.6105	$1.8396e^4$
Second order	HABF-APSO	0.2499	0.2580	0.3901	1.1500	$6.0364e^4$
	HABF-PSO	0.1721	0.2203	0.4612	0.4775	$6.7906e^4$
	PSO	0.3393	0.9008	0.3435	0.6318	$6.5545e^4$
	APSO	0.3044	0.3582	0.3350	1.1203	$6.0222e^4$
	ABFA	0.2645	0.4966	0.4047	0.6980	$6.3057e^4$

parameters of the fuzzy PID controller for both first and second order systems. These parameters are used for the various controllers compared subsequently. It is observed from the table that for the first order system, the proposed algorithm recorded the best error (J_{min}) of $1.3146e^4$. This is followed by ABFA with an error of $1.8396e^4$. For the second order system, the APSO algorithm produced the best error of $6.0222e^4$ followed closely by the proposed algorithm with $6.0364e^4$.

5.6.1 *First Order System*

Simulation results for the first order system with fuzzy PID controller optimised using various algorithms for step response are shown in Fig. 5.7. The zoom-in view of regions of interest A, B, C, and D are also shown in Fig. 5.8. Performance of the controller with various algorithms is summarised in Table 5.5. From both figures and table, it can be observed that the proposed HABF-APSO algorithm provides better setpoint tracking capability when compared to the other algorithms. Moderate response time with least overshoot (0.19%) and shortest settling time (25.11 s) are recorded for the HABF-APSO algorithm (see Table 5.5). The proposed HABF-APSO is a bit slower in response and recovery from disturbance effect as compared to APSO algorithm. However, this is achieved less overshoot as compared to latter. Observing the control signals, it is seen that the signal with proposed HABF-APSO is moderate at the point of step change and smoother at the point of disturbance when compared to other algorithms. As shown in the table, the IAEs for all controllers are within the range 11.836–13.8154. The best being that of APSO which is closely followed by the proposed HABF-APSO.

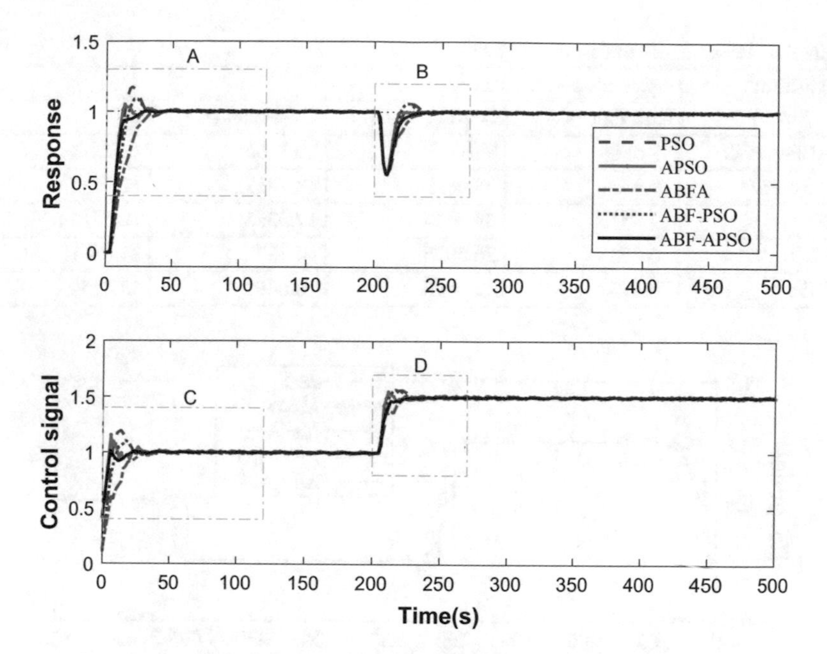

Fig. 5.7 Response of first order system with fuzzy PID tuned using various algorithms

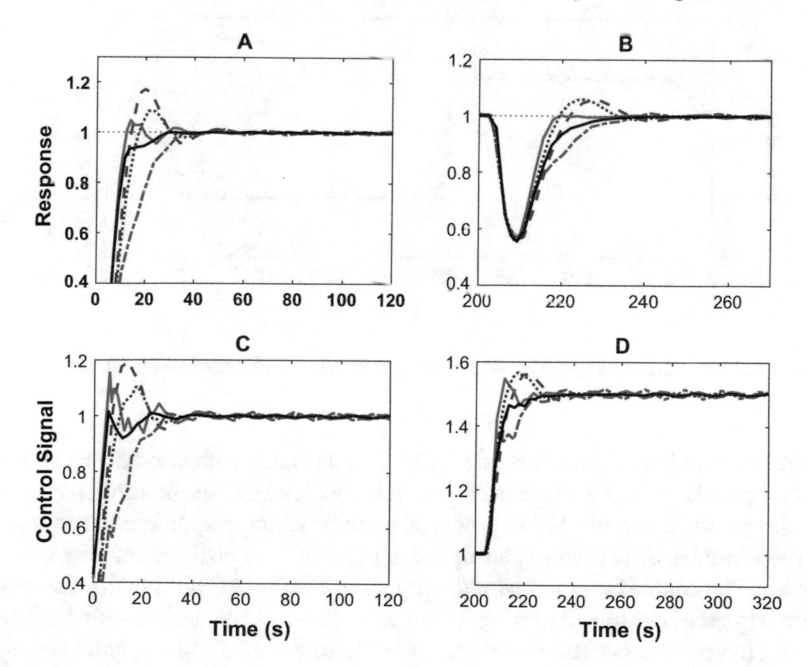

Fig. 5.8 Zoomed-in view of Fig. 5.7 for regions A, B, C and D

Table 5.5 Performance of first order system

Algorithm	System's performance			
	Rise time (s)	Settling time (s)	Overshoot (%)	IAE
HABF-APSO	7.5588	25.1130	0.1866	11.9273
HABF-PSO	9.4581	38.0695	8.1817	12.7453
PSO	7.5573	39.5414	17.3325	12.5514
APSO	6.3583	26.5095	4.1972	11.8361
ABFA	20.6952	32.4890	0.6400	13.8154

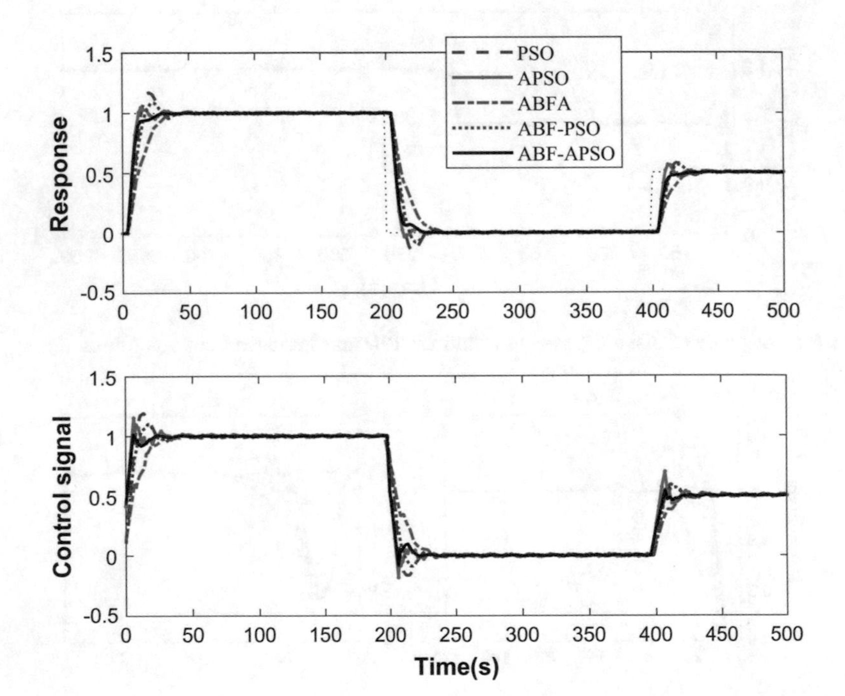

Fig. 5.9 Setpoint tracking response of first order system to fuzzy PID tuned with various algorithms

The variable setpoint tracking ability of the controller with various algorithms is also compared in Fig. 5.9. From the figure, it is observed that the controller optimised with the proposed HABF-APSO produced better tracking response compared to the other algorithms. This is more glaring at the point of both positive and negative step changes. The controller tuned with the proposed HABF-APSO produced smoother control signal as compared to when tuned with APSO, PSO, and HABF-PSO algorithms. However, the ABFA tuned controller despite producing a smooth signal, its response is too sluggish compared to when tuned with the proposed HABF-APSO algorithm.

5.6.2 Second Order System

In a similar way to the first order system, simulation results for the second order system with fuzzy PID controller optimised with various algorithms are shown in Fig. 5.10. The zoom-in view of regions of interest A, B, C, and D are also shown in Fig. 5.11. Furthermore, performance comparison of the controllers is given in Table 5.6. From the figures and table, it can be observed that the HABF-APSO algorithm outperformed the other algorithms in terms of settling time (36.43 s) and overshoot (0.77%). The controller tuned with the proposed algorithm also recovered from the disturbance effect no with no overshoot or oscillation as against the other algorithms. The response and control signal of the controller with proposed algorithm are both smoother than those with other algorithms. As shown in the table, the IAEs for all controllers are within the range 39.10–49.39. The best being that of HABF-PSO which is closely followed by the proposed HABF-APSO.

The changing setpoint tracking ability of the controller optimised with various algorithms for second order system is also compared in Fig. 5.12. From the figure, it is observed that the controller optimised with the proposed ABF-APSO algorithm produced response no overshoot or undershoot at the point of setpoint change when compared to the responses of the other algorithms. Generally, the response of the controller optimised with both HABF-APSO and HABF-PSO is smoother compared

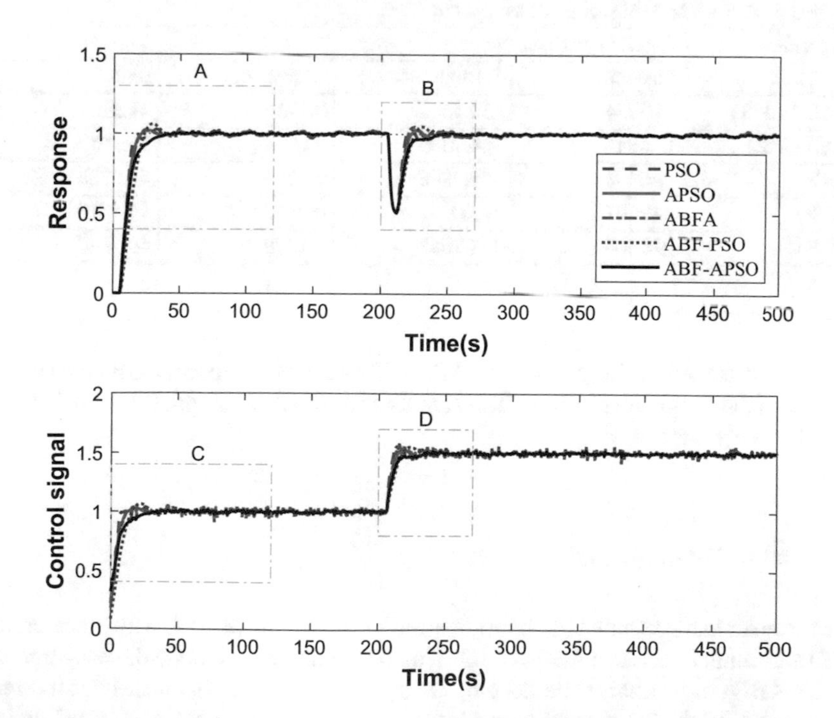

Fig. 5.10 Response of second order system to fuzzy PID tuned with various algorithms

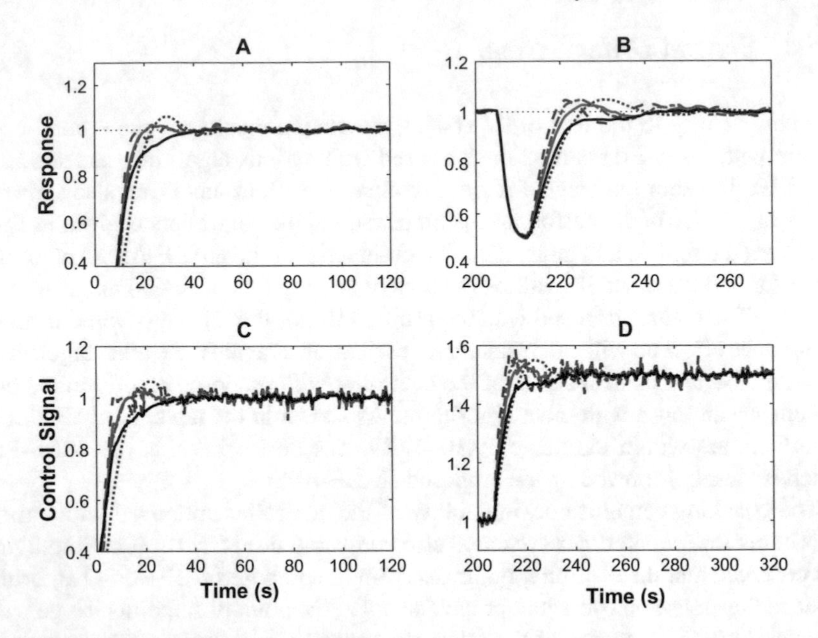

Fig. 5.11 Zoomed-in view of Fig. 5.10 for regions A, B, C and D

Table 5.6 Performance of second order system

Algorithm	System's performance			
	Rise time (s)	Settling time (s)	Overshoot (%)	IAE
HABF-APSO	14.3954	36.4318	0.7736	41.3204
HABF-PSO	12.4847	44.7455	5.7225	39.1057
PSO	7.0488	485.7308	2.9266	48.8758
APSO	9.2508	485.0259	0.9863	45.4616
ABFA	9.5500	484.7008	1.1688	49.3940

to the response with PSO, APSO, and ABFA. This can be corroborated by the control actions of the various controllers. The signals of the controller with PSO, APSO, and ABFA are oscillatory.

5.7 Summary

This chapter has presented the hybridization of the Adaptive BFA with Accelerated PSO algorithm to achieve ABF-APSO. This is in order to accelerate the convergence of the ABFA algorithm while keeping its dynamism. The effectiveness of the new proposed ABF-APSO algorithm can be seen in the benchmark function validation results. Also the application of optimal Fuzzy PID controller in a wireless networked

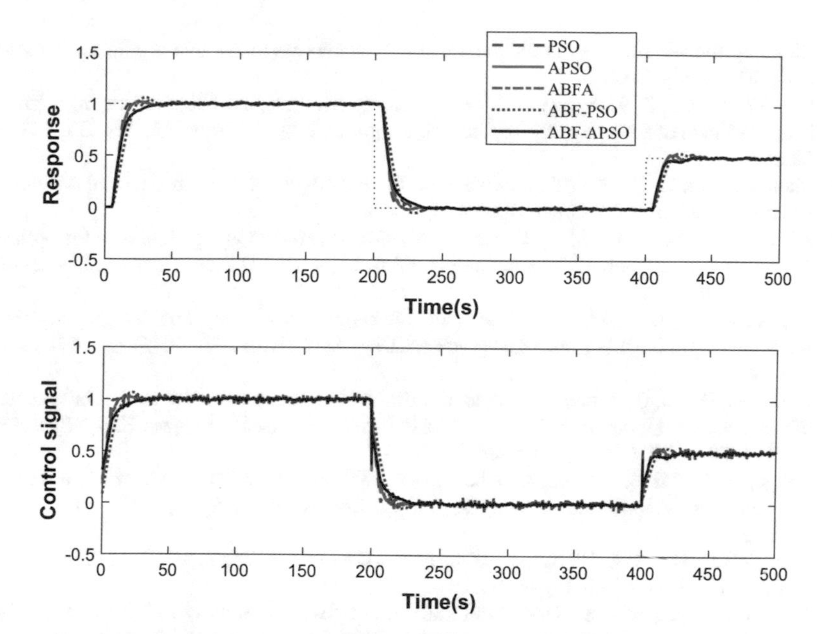

Fig. 5.12 Setpoint tracking response of second order system to fuzzy PID tuned with various algorithms

environment characterized by random variable networked induced delay and long deadtime plants. Furthermore, simulation results of the optimal Fuzzy PID controller tuned with the proposed algorithm (ABF-APSO) provides good settling time, overshoot and recovers fast from disturbance effect as compared to its performance with PSO, APSO, ABFA and ABF-PSO Algorithms.

References

1. Passino, K.M.: Biomimicry of bacterial foraging for distributed optimization and control. IEEE Control Syst. **22**(3), 52–67 (2002). IEEE
2. Panigrahi, B.K., Pandi, V.R.: Bacterial foraging optimisation: Nelder–Mead hybrid algorithm for economic load dispatch. IET Gener. Transm. Distrib. **2**(4), 556–565 (2008). IET
3. Panigrahi, B.K., Pandi, V.R.: Congestion management using adaptive bacterial foraging algorithm. Energy Convers. Manag. **50**(5), 1202–1209 (2009). Elsevier
4. Sathya, P.D., Kayalvizhi, R.: Optimal segmentation of brain MRI based on adaptive bacterial foraging algorithm. Neurocomputing **74**(14), 2299–2313 (2011). Elsevier
5. Kim, D.H., Abraham, A., Cho, J.H.: A hybrid genetic algorithm and bacterial foraging approach for global optimization. Inf. Sci. **177**(18), 3918–3937. Elsevier
6. Yıldız, Y.E., Altun, O.: Hybrid achievement oriented computational chemotaxis in bacterial foraging optimization: a comparative study on numerical benchmark. Soft Comput. **19**(2), 3647–3663 (2015). Springer
7. Hassan, S.M., Supriyono, H., Tokhi, M.O.: Direct control design with bacterial foraging algorithm. In: Adaptive Mobile Robotics, Proceedings of the 15th International Conference on

Climbing and Walking Robots and the Support Technologies for Mobile Machines, Baltimore, USA, 23–26 July 2012
8. Datta, T., Misra, I.S., Mangaraj, B.B., Imtiaj, S.: Improved adaptive bacteria foraging algorithm in optimization of antenna array for faster convergence. Prog. Electromagn. Res. C **1**, 143–157 (2008)
9. Chen, H., Zhu, Y., Hu, K.: Adaptive bacterial foraging optimization. Abstract and Application Analysis. Hindawi Publishing Corporation, London (2011)
10. Farhat, I.A., El-Hawary, M.E.: Dynamic adaptive bacterial foraging algorithm for optimum economic dispatch with valve-point effects and wind power. IET Gener. Transm. Distrib. **4**(9), 989–999 (2010). IET
11. Supriyono, H., Tokhi, M.O.: Parametric modelling approach using bacterial foraging algorithms for modelling of flexible manipulator systems. Eng. Appl. Artif. Intell. **25**(5), 898–916 (2012). Elsevier
12. Majhi, R., Panda, G., Majhi, B., Sahoo, G.: Efficient prediction of stock market indices using adaptive bacterial foraging optimization (ABFO) and BFO based techniques. Expert Syst. Appl. **36**(6), 10097–10104 (2009). Elsevier
13. Dasgupta, S., Das, S., Biswas, A., Abraham, A.: Automatic circle detection on digital images with an adaptive bacterial foraging algorithm. Soft Comput. **14**(11), 1151–1164 (2010). Springer
14. Kim, D.H.: Hybrid GA–BF based intelligent PID controller tuning for AVR system. Appl. Soft Comput. **11**(1), 11–12 (2011). Elsevier
15. Biswas, A., Dasgupta, S., Das, S., Abraham, A.: Synergy of PSO and bacterial foraging optimization—a comparative study on numerical benchmarks. Innov. Hybrid Intell. Syst. 255–263 (2007). Springer
16. Nasir, A.N.K., Tokhi, M.O.: A novel hybrid bacteria-chemotaxis spiral-dynamic algorithm with application to modelling of flexible systems. Eng. Appl. Artif. Intell. **33**, 31–46 (2014). Elsevier
17. Nasir, A.N.K., Tokhi, M.O.: An improved spiral dynamic optimization algorithm with engineering application. IEEE Trans. Syst. Man Cybern.: Syst. **45**(6), 943–954 (2015)
18. Hota, P.K., Barisal, A.K., Chakrabarti, R.: Economic emission load dispatch through fuzzy based bacterial foraging algorithm. Int. J. Electr. Power Energy Syst. **32**(7), 794–803 (2010). Elsevier
19. Pan, I., Das, S., Gupta, A.: Tuning of an optimal fuzzy PID controller with stochastic algorithms for networked control systems with random time delay. ISA Trans. **50**(1), 28–36 (2011). Elsevier
20. Yang, X.S.: Engineering Optimization: An Introduction with Metaheuristic Applications. Wiley, New York (2010)

Part III
Comparative Study of Various Controllers on WirelessHART Networked Systems

Chapter 6
Comparison on WirelessHART Networked Systems

6.1 Introduction

In this chapter, comparison of the performance of the proposed control approaches is given for both simulation and practical applications. Here, three of the proposed controllers namely, SW, FPPI and Fuzzy PID are compared alongside conventional PI and Smith Predictor controllers. In both simulation and practical cases, the FPPI and Fuzzy PID are optimised using one of our proposed algorithm; the HABF-APSO.

The First section of the chapter will present comparison in the simulation environment. This comparison will be a second order system. The next section of the chapter will compare controllers in the practical environment. The controllers will be tested on a Flow Control and Calibration Process Mobile Pilot plant. Lastly, a summary on all the controllers will be provided.

6.2 Comparison in the Simulation Environment

The controllers here are compared based on their ability to track setpoint and respond to abrupt disruption (disturbance) under influence of stochastic network delay and noise. This comparison is done for two scenarios. In the first instance, the performance of various controllers is compared for step input signal with disturbance. In the second instance, the ability to track changing setpoint of these controllers is compared. For the purpose of this comparison, the second order time delay system is selected.

6.2.1 Performance of Various Controllers to Step Reference

The performance comparison of the proposed SW, FPPI and Fuzzy PID controllers alongside PI and Smith Predictor for step signal is given in Fig. 6.1. As usual, the regions of interest (i.e. point of step change and disturbance) for both response and

S. M. Hassan et al., *Hybrid PID Based Predictive Control Strategies for WirelessHART Networked Control Systems*, Studies in Systems, Decision and Control 293,
https://doi.org/10.1007/978-3-030-47737-0_6

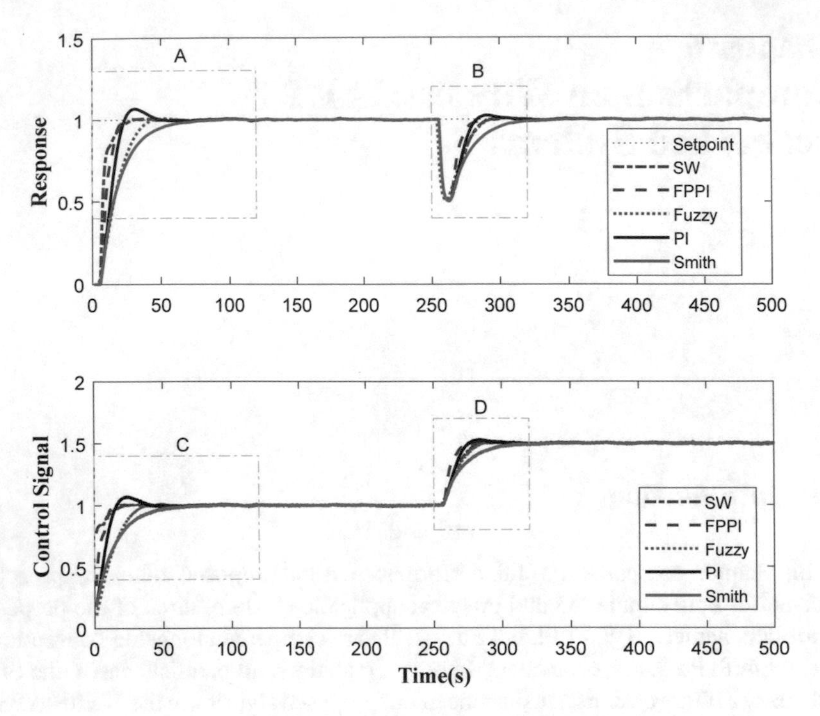

Fig. 6.1 Second order system response to SW, FPPI Fuzzy PID and PI controllers

control signals are highlighted in Fig. 6.2. The numerical results from these figures are presented in Table 6.1. From the result in the table, respective overshoots of the SW, FPPI, Fuzzy PID, PI and Smith Predictor are given as 1.04, 0.94, 1.08, 7.21 and 0.52%. It is noted that the Smith Predictor produced the least overshoot while all proposed controllers have overshoots around 1%. This is an improvement on 7% of the PI controller. At approximately 10 s each, the proposed SW and FPPI have faster rise times compared to 12.7, 24.2 and 31.4 s of the PI, Fuzzy PID and Smith predictor. However, the FPPI settled faster at 21.9 s followed by SW at 24.1 s. The Fuzzy PID, PI, and Smith Predictor also have their respective settling time as 39.2, 44.7 and 62.1 s. From the rise and settling times results, it can be seen that the Smith Predictor is the most sluggish despite having least overshoot. On the recovery from disturbance effect as observed from Fig. 6.2, the FPPI is the fastest to recover without overshoot. While the PI controller shows faster recovery than SW, Fuzzy PID, and Smith predictor, it does that at the expense of some overshoot. Furthermore, as observed previously, the Smith Predictor is the last to recover from the disturbance. The IAE of the controllers also followed a similar pattern to the rise time. Consequently, the value for both SW and FPPI is approximately 176, while PI, Fuzzy PID and Smith Predictor have approximate values of 221, 263 and 306 respectively. The signals generated by the compared controllers also shows that during step change, higher signals starting at around 0.4 and 0.8 are generated by the

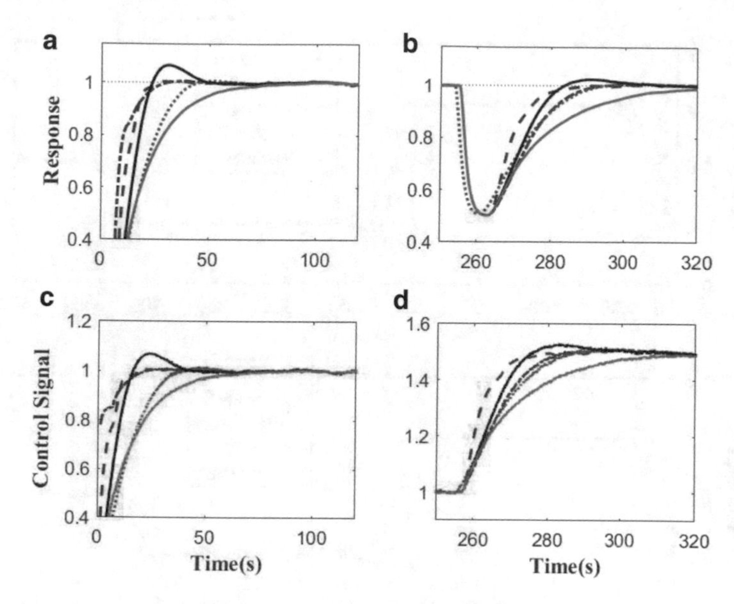

Fig. 6.2 Zoomed-in view of Fig. 6.2 for regions A, B, C and D

Table 6.1 Performance of controllers for second order system

Controller	System's performance			
	Rise time (s)	Settling time (s)	Overshoot (%)	IAE
SW	10.0123	24.1357	1.0374	176.1628
FPPI	10.1006	21.9377	0.9450	175.8783
Fuzzy	24.2278	39.1881	1.0779	262.6770
PI	12.7245	44.6820	7.2109	221.3465
Smith	31.4080	62.1434	0.5182	305.7985

FPPI and SW compared to the 0.1–0.15 of the other controllers. This is responsible for the faster response of the two controllers.

The tracking performance of the controllers is compared by simulating the second order system to changing setpoint signal and the result is shown in Fig. 6.3. From the figure, it can be clearly seen that the FPPI and SW controllers ability for tracking changing setpoint outperformed those of the other controllers. This is in conformity with the results obtained for the step response.

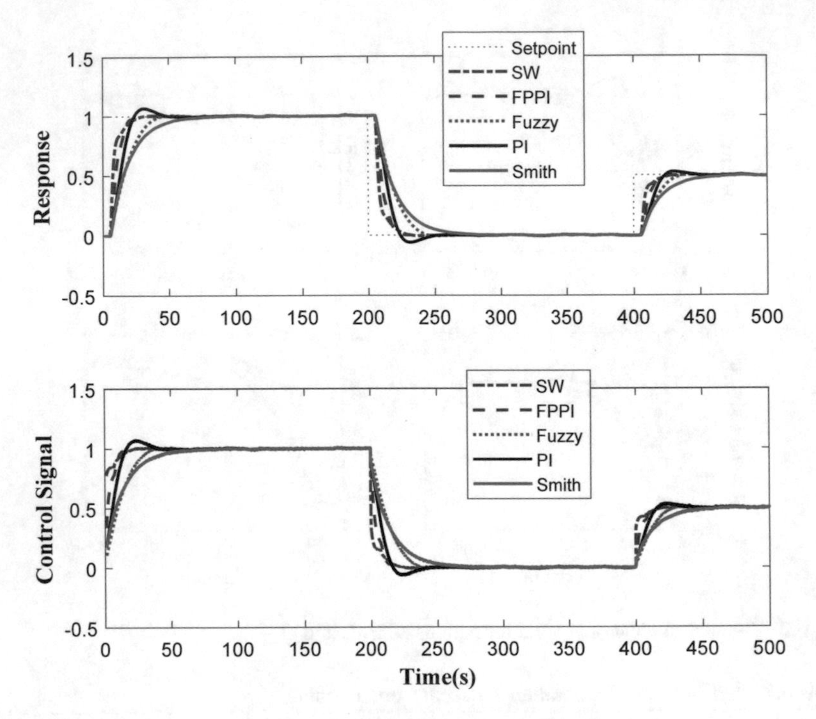

Fig. 6.3 Changing reference tracking of SW, FPPI and Fuzzy controller

6.3 Comparison in the Practical Environment

The controllers are compared here based on their ability to track setpoint under real-time network conditions. The experimental set-up discussed in Sect. 6.3.1 is used. In this section, the various controller parameters will be given first. Then, experimental result will be presented afterwards.

6.3.1 Implementation on Flow Control and Calibration Process Pilot Plant

The developed controllers will be implemented on a *PcA SimExpert* Flow Control and Calibration Process Mobile Pilot plant stationed at Block 23, Universiti Teknologi PETRONAS. The complete plant diagram and the simplified P&ID are shown in Figs. 6.4 and 6.5 respectively. The plant consist of a buffer tank and a calibration tank that are connected in series. The objective of the plant is to transfer fluid from the buffer tank to the calibration tank at a controlled flow rate. To achieve this, two pumps and a pneumatic valve are used. In this experiment, the measured value is water level while the manipulated value is the flow rate in m^3/s. From Fig. 6.5, PIC110C is

Fig. 6.4 Pilot flow control pilot plant

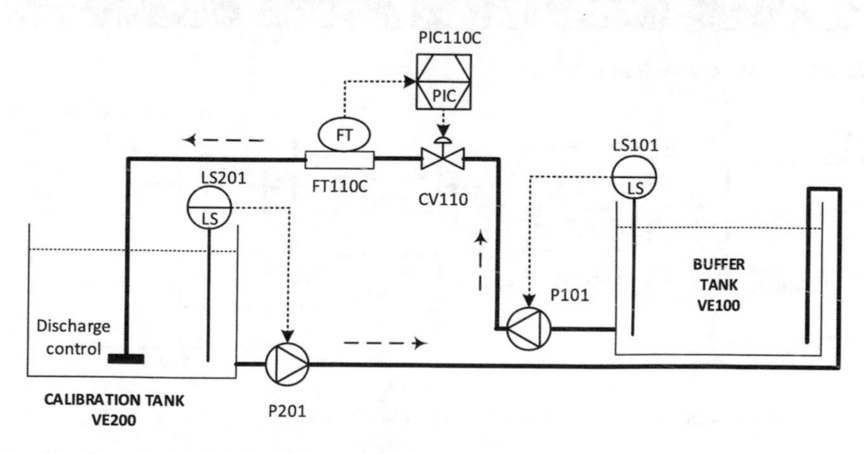

Fig. 6.5 Simplified P&ID diagram of the pilot flow control plant

the main feedback controller while FT110C and CV110 are the flow transmitter and the control valve respectively. P101 and P201 are the pumps for buffer tank VE100 and calibration tank VE200 respectively. In the same vein, LS101 and LS201 are the respective level sensors connected to P101 and P201 for the control of overflow.

The model of the system is given in the transfer function of (6.1).

$$G = \frac{0.58}{0.26s + 1} e^{-0.25\,s} \tag{6.1}$$

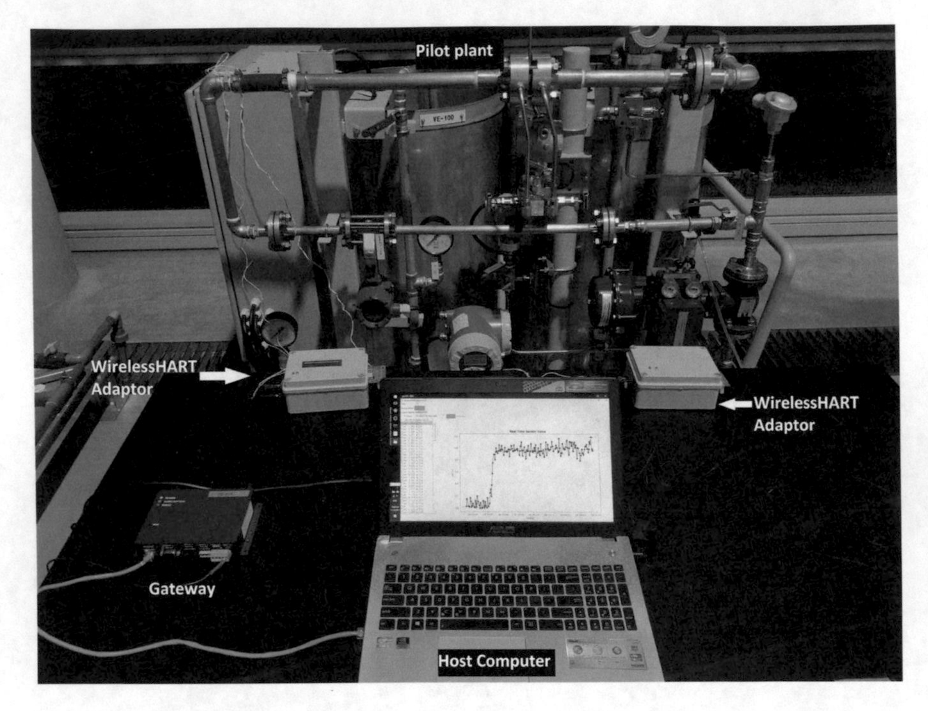

Fig. 6.6 Complete experimental set-up

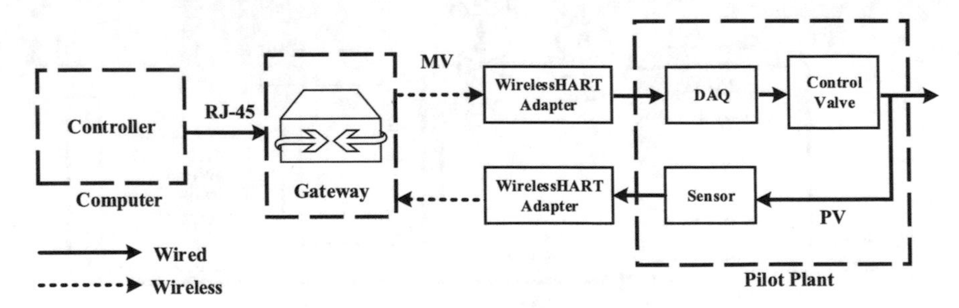

Fig. 6.7 Block diagram representation of the experimental set-up

The complete experimental set-up is shown in Fig. 6.6 while the block diagram representation is given in Fig. 6.7. To achieve wireless control of the flow rate by controlling the opening of CV110 and to obtain flow measurement, the controller is implemented in Simulink environment in the host computer interfaced with Python to export the control action into the gateway. The control signal or manipulated variable (MV) is now received by the valve through a developed WirelessHART adapter. Measurements based on the process variable (PV) are received from the sensor via WirelessHART adaptor into the gateway and then the controller.

Table 6.2 Various controller parameters

Controller	Parameters
PI/Smith predictor	$K_c = 1.2354$, $K_i = 3.8801$
FPPI	$T_f = 0.2041$, $T_i = 0.26$, $K_c = 1.2334$
SW	$G_r = 20$, $G_{yr} = \frac{1}{0.26s+1}$, $K_c = 0.0996$, $K_i = 0.383$
Fuzzy PID	$K_e = 0.3011$, $K_{ce} = 0.1222$, $\alpha = 0.8465$, $\beta = 8269$

6.3.2 Controller Parameters

In order to practically validate the developed control strategies, the controllers were tested on a pilot flow control system of Fig. 6.4. The various controller parameters for the plant are given in Table 6.2. The same PI controller parameters are used for the PI in the Smith Predictor set-up. Similar to other sections, the Fuzzy PID and FPPI parameters were selected through tuning with the proposed HABF-APSO algorithm. The optimisation was done using model of the plant given in Eq. (6.1).

6.3.3 Experimental Result

In this experiment, the data sampling time in MATLAB is set as 0.5 s while the update rate between the gateway and field devices (motes/adaptors) is set as the 4 s which is at the middle of 1 s for fast piping and 8 s for optimal battery performance. The disadvantage of the fast piping is that it drains the batteries of field devices faster. The duration of the experiment for each controller tested is 500 s. This is due to the capacity of the tanks VE100 and VE200. The target was set at 1.5 m^3/h. It should be noted that the valve position at 100% delivers a maximum of 2.5 m^3/h.

Result comparison of the PI, FPPI, SW, Fuzzy PID and Smith Predictor controllers is shown in Fig. 6.8. The numerical information of the result as obtained form the figure is given in Table 6.3. From the figure, it can be clearly seen that FPPI controller has faster response with a rise time of around 34 s. This is followed by PI and SW with 46.77 s and 56.22 s each. The slowest are the fuzzy and Smith Predictor with respective rise times of 72.27 and 164.73 s. Although the Smith Predictor recorded the least overshoot of 6.7%, it is however very slow in response and experiences glitch at around 150s during the experiment. Comparing overshoots of the proposed controllers against the PI, it can be seen that there is a significant difference between the 31.33% of the PI and the 11.33, 14.67 and 16% of the SW, Fuzzy and FPPI respectively. In a nutshell, all the proposed controllers have outperformed the PI in terms of overshoot while the FPPI has an added advantage of rise time over the PI.

The respective characteristics of each controller as seen from its response is also manifested through the control signals. While the signal of FPPI rose steadily to 40% in less than 50 s, the Smith Predictor is the most sluggish by reaching the 40%

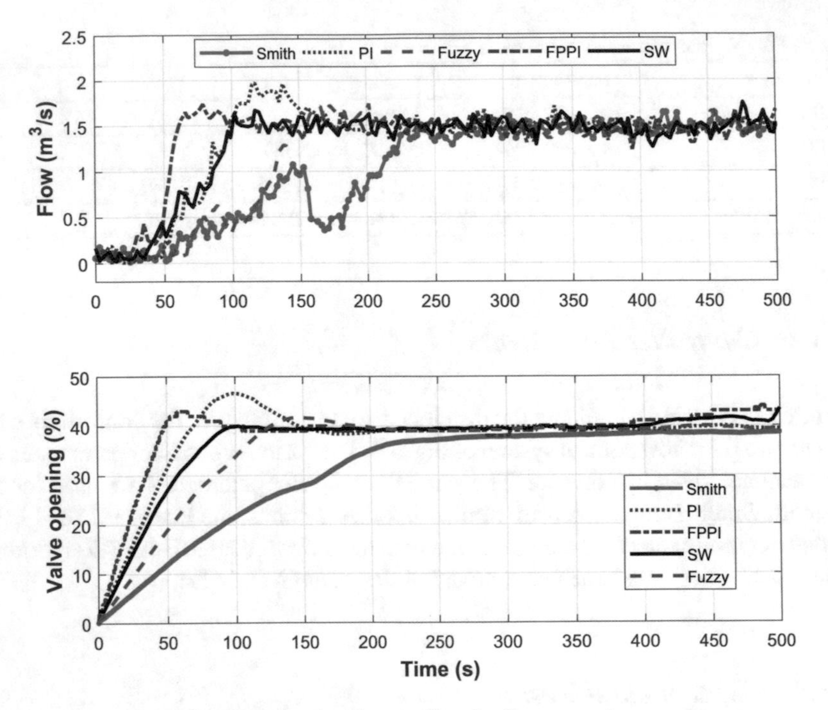

Fig. 6.8 Comparison of various developed controllers for flow control

Table 6.3 Performance of controllers for pilot plant

Controller	System's performance	
	Rise Time (s)	Overshoot (%)
SW	56.22	11.33
FPPI	33.95	16.00
Fuzzy PID	72.27	14.67
PI	46.77	31.33
Smith pred.	164.73	6.70

mark at around 200 s. It should also be noted that, the signal of SW despite being slower than both FPPI and PI, does not go beyond the 40% position of the valve as compared to the FPPI, PI and Fuzzy PID. This is responsible for its lower overshoot compared to the three. The pattern of results obtained here reflect the one observed for the simulation. This is to say that, the proposed controllers have improved on both the PI and Smith Predictor controllers.

6.4 Summary

The chapter has presented and discussed results of the SW, FPPI and Fuzzy PID control strategies proposed for WHNCS. The goal of these strategies all being PID based is to improve the performance of the PID while retaining its simple structure. The SW has been compared against PI and Smith Predictor (also PID based). For the case of wide variation in network delay, the FASW has been compared SW and PI. Meanwhile, the FPPI controller has been compared against the PI, ordinary PPI and Smith predictor. Furthermore, the performance of the proposed HABF-APSO algorithm has also been evaluated against its constituent algorithms on the Fuzzy PID algorithm. The comparisons and performance evaluation are done in both simulation and practical environments for all controllers. Therefore, the findings in the result section can be summarised as follows:

In both Simulation and practical applications, the SW strategy has generally improved the tracking performance of both PI and Smith Predictor controllers. This is in-terms of faster rise and settling times and recovery from disturbance effect. Furthermore, the use of SW on the PI has also drastically reduced its overshoot.

It should be noted that in case of wider variation of the network delays where the SW could not provide the needed good setpoint tracking, FASW is employed. The FASW is able to adjust the SW controller to maintain its good tracking ability. This is even more evident for higher order systems.

The FPPI controller has also improved the performance of PPI especially for higher order systems and on its robustness to measurement noise. Furthermore, similar to the SW approach, this controller has outperformed the PI and Smith Predictor in terms of setpoint tracking and disturbance regulation.

The proposed HABF-APSO algorithm has been compared with ABFA, APSO, PSO and HABF-PSO algorithms on the Fuzzy PID controller. Performance of the controller tuned with the proposed algorithm shows faster settling time and lower overshoot as compared to other algorithms. Hence, better tracking performance of the controller is achieved.

A general comparison of all the proposed controllers showed that these controllers have improved on the setpoint tracking performance of both PI and Smith Predictor approaches for both simulation and practical applications.

Index

S. M. Hassan et al., *Hybrid PID Based Predictive Control Strategies for WirelessHART Networked Control Systems*, Studies in Systems, Decision and Control 293,
https://doi.org/10.1007/978-3-030-47737-0